Advances in Computing Communications and Informatics
(Volume 2)
Smart Antennas: Recent Trends in Design and Applications

Edited by

Praveen Kumar Malik
School of Electronics and Electrical Engineering
Lovely Professional University
Punjab,
India

Pradeep Kumar
University of KwaZulu-Natal
Durban-4041,
South Africa

Sachin Kumar
Amity University Lucknow
Uttar Pradesh,
India

&

Dushyant Kumar Singh
School of Electronics and Electrical Engineering
Lovely Professional University
Punjab,
India

Advances in Computing Communications and Informatics

Smart Antennas: Recent Trends in Design and Applications

Volume # 2

Editors: Praveen Kumar Malik, Pradeep Kumar, Sachin Kumar and Dushyant Singh

ISSN (Online): 2737-5730

ISSN (Print): 2737-5722

ISBN (Online): 978-1-68108-859-4

ISBN (Print): 978-1-68108-860-0

ISBN (Paperback): 978-1-68108-861-7

need for a court order if at any point you breach any terms of this License Agreement. In no event will any delay or failure by Bentham Science Publishers in enforcing your compliance with this License Agreement constitute a waiver of any of its rights.

3. You acknowledge that you have read this License Agreement, and agree to be bound by its terms and conditions. To the extent that any other terms and conditions presented on any website of Bentham Science Publishers conflict with, or are inconsistent with, the terms and conditions set out in this License Agreement, you acknowledge that the terms and conditions set out in this License Agreement shall prevail.

Bentham Science Publishers Ltd.
Executive Suite Y - 2
PO Box 7917, Saif Zone
Sharjah, U.A.E.
Email: subscriptions@benthamscience.net

BENTHAM SCIENCE

CONTENTS

PREFACE

The book primarily focuses on recent advances in the field of micro-strip antenna design and its applications in various fields, including space communication, mobile communication, wireless communication, medical implants, wearable applications *etc*. Scholars from electronics/ electrical/instrumentation engineering, researchers, and industrial people will benefit from this book. Students, researchers, and people from industries have expressed concerns about obtaining antenna measurements in various environments. The current book shall provide the literature using which students and researchers would be able to design antennas for above-mentioned applications. Ultimately, it would enable users to take measurements in different environments. The book is organized into eight chapters. A brief description of each chapter is as follows:

Introduction to Microstrip Antenna: The chapter includes basics of microstrip antenna, its design issues, and various applications.

Fractal and Defected Ground Structure Microstrip Antenna for Wireless Applications: The chapter highlights the characteristics of antenna miniaturization technology across distinct fractal geometries. Non-Euclidian geometry is fractal geometry. The fractal geometry antenna has multi-band frequency of operation with a relatively larger bandwidth. Fractal antennas have demerits that can be effectively optimized by defective ground structure over traditional antennas. Numerous applications such as WLAN, Wi-Fi, Bluetooth, WCN, Mobile 2G,3G, 4G, 5G, Wi-Max, *etc*., are seen in the research discussed in this article.

Role of AI in Market Developments and Financial Stability Implications: This research article provides a general role of AI in financial services and its definition in practical scenarios. The research work is mostly qualitative and does not include any data analysis for its results. This chapter describes the major benefits derived by the practices of AI in general in the financial sector. For the study, some examples of the role of AI in service organizations are analyzed and results are given from their findings. The authors have explained the benefits of AI from the perspectives of both customers and service providers.

Performance Analysis of Microstrip Patch Antenna for Various Applications: With the development of wireless communication systems, compact wireless devices that allow more space for the integration of other electronic components are needed. Engineering innovation poses problems for integrating multiple RF-band antennas with a wide range of frequencies. We can improve the antenna design by advancing the optimization methodology. It also provides us with the opportunity to analyze the existing studies to categorize and synthesize them in a meaningful way.

Importance and Uses of Microstrip Antenna in IoT: This chapter attempts to consider how the IoT (Internet of Things) is revolutionizing the world. IoT is a network similar to the monster wherein different devices interconnect and furnish to communicate with each other. Thus, it helps to drive computerization to an advanced level. This helps all the connected devices to communicate with one another and make decisions on their own without human intercessions.

Use of Smart-Antenna in Mobile Communication: By researching the smart antenna techniques, it is observed that the success of the smart antenna lies on two major factors: first, the smart antenna's features need to be considered early in the design phase of future systems (top-down compatibility); second, a realistic performance evaluation of smart antenna techniques needs to be performed according to the critical parameters associated with future system requirements (bottom-up feasibility). In the end, we will discuss the market trends, future projections, and the expected financial impact of smart antenna systems deployment.

Praveen Kumar Malik
School of Electronics and Electrical Engineering
Lovely Professional University
Punjab
India

Pradeep Kumar
University of KwaZulu-Natal
Durban-4041
South Africa

Sachin Kumar
Amity University Lucknow
Uttar Pradesh
India

&

Dushyant Kumar Singh
School of Electronics and Electrical Engineering
Lovely Professional University
Punjab
India

DEDICATION

This book is dedicated to my late father, who taught me to be an independent and determined person, without whom I would never be able to achieve my objectives and succeed in life.

Late (Sr.) Dharamveer Singh

List of Contributors

Abdul Rahim	Lovely Professional University, Punjab, India
Arghya Majumder	Department of Electrical and Electronics Engineering, Lovely Professional University, Punjab, India
Arshi Naim	Department of Information Systems, College of Computer Science, King Khalid University, Abha 62529, Saudi Arabia
Charu Tyagi	Department of Electronics and Communication Engineering, Raj Kumar Goel Institute of Technology, Ghaziabad, India
Fahad Alahmari	Department of Information Systems, College of Computer Science, King Khalid University, Abha 62529, Saudi Arabia
Kiran Srivastava	Department of Electrical and Electronics Engineering, Galgotias College of Engineering and Technology, Greater Noida, India
Patrika Jayanti	Department of Electronics Engineering Mahamaya Polytechnic of IT, Hathras, Uttar Pradesh 202002, India
Puneet Chandra Srivastava	Department of Electronics and Communication Engineering, Raj Kumar Goel Institute of Technology, Ghaziabad, India
R. Nagarajan	Department of Electrical and Electronics Engineering, Gnanamani College of Technology, Namakkal, Tamil Nadu, India
S. Kannadhasan	Department of Electronics and Communication Engineering, Cheran College of Engineering, Anna University, Tamil Nadu, India
Vivek Arya	Department of Electronics Communication & Engineering, Faculty Of Engineering & Technology, Gurukul Kangri Vishwavidyalaya Haridwar, Uttarakhand 249404, India
Wani V. Patil	Department of Electronics Engineering, G. H. Raisoni College of Engineering, Nagpur, Maharashtra 440016, India

CHAPTER 1

Introduction to Microstrip Antenna

Vivek Arya[1,*]

[1] *Department of Electronics Communication & Engineering, Faculty Of Engineering & Technology, Gurukul Kangri Vishwavidyalaya Haridwar, Uttarakhand 249404, India*

Abstract: The very first idea of microstrip antenna was given by G. A. Deschamps in 1953 [1]. However, it did not receive practical exposure until the 1970s and it was further developed by Robert E. Menson [2]. The microstrip antennas are also called *patch* antennas and abbreviated as MSA. The microstrip antenna has various key advantages due to its low profile, light weight, low cost, and miniaturization capability [3-4]. There are various authentic applications of microstrip antenna such as satellite communication, Radar, WLAN, and WiMAX [5-9]. Nowadays, microstrip antennas are widely used for military and civilian applications such as broadcast radio, television, mobile systems, radio-frequency identification (RFID) system, vehicle guidance system, a global positioning system (GPS), vehicle collision avoidance system, multiple-input multiple-output (MIMO) systems, radar systems, determination of direction, surveillance systems, biological imaging, and missile systems, *etc.*

Keywords: Directivity, Gain, Patch antenna, Radiation pattern.

1. INTRODUCTION

Several researchers and experts are working to improve the various quality parameters like bandwidth, directivity, and gain of microstrip antenna. Some other existing solutions, such as defected ground structures (DGS), electromagnetic bandgap (EBG) structures, and composite resonator structures, create the issues of spurious radiation and very high complexity. The new approach provides the solution for this problem using metamaterial. In 1968, Russian Physicist Prof. Vaselago was the first who theoretically proposed the concept of the metamaterial. Attractive and interesting properties of metamaterials play a very important and authentic role in antenna designing. Therefore, the metamaterial can be used for the performance enhancement of microstrip patch antennas, as shown in Fig. (**1**).

* **Corresponding author Vivek Arya:** Department of ECE, FET, Gurukul Kangri Vishwavidyalaya Haridwar, India.
E-mail: ichvivekmalik@gmail.com

Praveen Kumar Malik, Pradeep Kumar, Sachin Kumar and Dushyant Kumar Singh (Eds.)

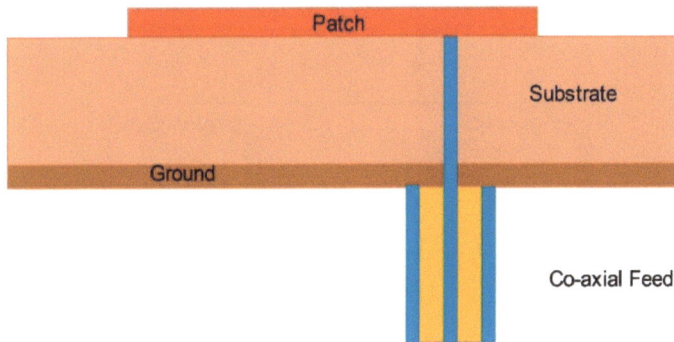

Fig. (1). Geometrical Layout of Microstrip Antenna, 3-Dimensional view of microstrip antenna.

2. FEEDING TECHNIQUES

The main objective of the feedline is to provide an input signal to the antenna for excitation. Nowadays, several feeding techniques are available for microstrip patch antennas. These feeding techniques or methods are categorized into two groups (a) contacting and (b) non-contacting technique, as shown in Fig. (2). In the contact technique, using connecting elements (like a coaxial probe, microstrip line, and inset fed or notch fed), the RF power is directly fed to the patch of an antenna. In non-contacting techniques, various feed methods, *i.e..*, proximity coupling, aperture coupling, and electromagnetic field coupling, are used for power transfer. The feeding technique selection for microstrip antenna plays a very crucial role because various antenna quality parameters such as return loss, bandwidth, and efficiency of an antenna are directly affected by it [10]. The surface waves and spurious feed radiation vary according to the thickness of the substrate that restricts the bandwidth of the microstrip antenna [11]. The co-axial feed and microstrip feed contacting techniques are the most commonly used in patch antenna designing, while in the non-contacting feed technique, the radiating patch is indirectly fed by the RF power, and then power is transferred to radiating patch. The most popular and commonly used techniques in non-contacting methods are aperture coupled feed and proximity coupled feed [12 - 14]. Comparison of different feeding techniques is given in Table **1**. Summary of advantages and disadvantages of different feeding techniques are given in Table **2**. The general feeding techniques are discussed briefly as follows.

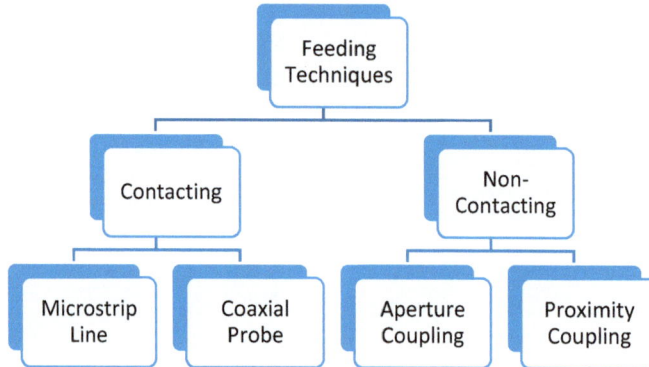

Fig. (2). Classification of different feeding techniques.

Table 1. Comparison between different existing feeding methods.

Characteristics	Microstrip Line Feed	Coaxial Feed	Aperture Coupled Feed	Proximity Coupled Feed
Spurious Feed Radiation	Higher	Lower	Moderate	Lowest
Reliability	Excellent	Due to Poor Soldering	Very Good	Very Good
Ease of Fabrication	Simple and Easy	Soldering and Drilling needed	Alignment Required (Very Difficult)	Alignment Required
Impedance Matching	Simple and Easy	Simple and Easy	Simple and Easy	Simple and Easy
Bandwidth	Narrow (2-5%)	Narrow (2-5%)	Narrow (2-5%)	Wider (13%)

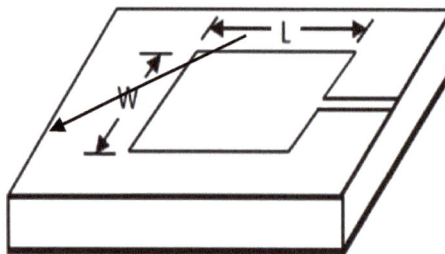

Fig. (3). Microstrip line feed for microstrip antenna.

i). *Microstrip Feedline*: A metallic strip is directly linked with a patch, as shown in Fig. (**3**). In feed arrangement, the feed and patch are grafted on the same substrate, due to which it becomes a flat structure [15 - 16]. Its impedance matching and fabrication are easy. It produces the highest spurious feed radiation compared to other feeding techniques and provides a narrow bandwidth (2% to 5%).

ii). *Coaxial Feed*: In literature, the coaxial feed is also called by the name probe feed. In this feeding method, the inner conductor of the coaxial is connected with the radiating patch, and the outer conductor is attached to the ground plane [17]. It produces lower spurious radiations and requires simple and easy impedance matching. The coaxial feed arrangement is made with the help of soldering and drilling. It also provides narrow bandwidth of nearly 2 to 5%.

iii) *Aperture Coupled Feed*: In aperture coupled feed, two different layers or substrates are sandwiched to obtain the electromagnetic coupling from feed to radiating patch of microstrip antenna [18]. It has moderate spurious radiation and easy impedance matching. The aperture coupled feed arrangement is very difficult due to alignment requirements. This technique provides narrow bandwidth of nearly 2 to 5%.

iv) *Proximity Coupled Feed*: The feed line are sandwiched between the two dielectric layers, and the radiating patch is kept on the top of the upper dielectric substrate [19]. This method of feeding removes the spurious feed radiation. This feeding technique provides wider bandwidth (nearly 13%).

Table 2. List of Advantages and Disadvantages of different feeding methods.

Feeding Techniques	Advantages	Disadvantages
Coaxial Feed	➤ Simple and easy impedance matching ➤ It has lower spurious radiations	➤ It has large inductance for thick substrate ➤ For it, soldering is needed
Microstrip Line	➤ It is monolithic ➤ It is simple and easier to fabricate ➤ Simple and easy to impedance matching	➤ For thick substrate, it produces spurious radiation
Aperture Coupled	➤ It does not have a deleterious effect on the bandwidth and efficiency due to the use of two dielectric layers. ➤ It does not require direct contact between feed and patch ➤ It does not produce radiation from feed and other active devices	➤ It required a multi-layer design ➤ It produces higher radiation from the back lobe.
Proximity Coupled Feed	➤ Easy impedance matching ➤ Provide wider bandwidth	➤ Fabrication needed alignment ➤ Less spurious fed radiation

3. DESIGNING OF MICROSTRIP ANTENNA

To design the rectangular microstrip patch antenna, some parameters and physical dimensions of the patch are calculated as follows [20 - 23]:

Step 1: Choose a dielectric constant for the substrate and resonant frequency.

Step 2: Compute the width using equation (1)

$$W = \frac{c}{2f_r}\sqrt{\frac{2}{\varepsilon_r + 1}} \tag{1}$$

where W is the width of the patch, c is the speed of light E_r, is the dielectric constant

Step 3: Now, find the value of the dielectric constant using equation (2)

$$\varepsilon_{reff} = \frac{\varepsilon_r + 1}{2} + \frac{\varepsilon_r - 1}{2}[1 + 12\frac{h}{w}]^{-\frac{1}{2}} \tag{2}$$

Step 4: Increment in length (ΔL) computed by equation (3)

$$\frac{\Delta L}{h} = 0.412\frac{(\varepsilon_{reff} + 0.3)(\frac{w}{h} + 0.264)}{(\varepsilon_{reff} - 0.258)(\frac{w}{h} + 0.8)} \tag{3}$$

where, h is the height of substrate.

Step 5: Now, compute the Length (L) using equation (4)

$$L = \frac{c}{2f_r\sqrt{\varepsilon_{reff}}} - 2\Delta L \tag{4}$$

Step 6: Determination of dimension of the ground by

$$\left.\begin{array}{l} L_g = L + 6h \\ W_g = W + 6h \end{array}\right\} \tag{5}$$

where L_g and W_g are the length and width of the ground, respectively

So, by carefully following the above-mentioned six steps, one can design the microstrip patch antenna of any desired resonant frequency.

Note: The size and frequency of a microstrip patch antenna are inversely proportional to each other.

4. ADVANTAGES AND DISADVANTAGES OF MICROSTRIP ANTENNA ARE SUMMARIZED AS BELOW

Advantages:

a) It is compact in size.

b) It has very little weight.

c) It is less expensive due to the low cost of fabrication.

d) Its installation is easy and very simple.

e) Its fabrication process is very easy and simple.

f) It shows very good performance.

g) Easy to design a large array antenna, which can be spaced at half-wavelength or less.

h) By this, conformal designs are possible because it is very easy to create curved surfaces.

i) It does not require cavity backing.

j) It provide linear and circular polarization.

k) Dual and triple frequency operations are possible.

Disadvantages:

a) It has poor scanning capability.

b) It shows poor polarization purity.

c) It provides spurious feed radiation

d) It gives narrow frequency bandwidth (*i.e.*, 1 to 5%, but it can be increased by increasing complexity).

e) It provides less power so, it has low power handling capability.

f) It has low efficiency.

g) It provide low gain.

h) It produces ohmic losses in the feeding arrangement of array structures.

5. APPLICATIONS OF MICROSTRIP ANTENNA

The microstrip patch antennas are popular due to their simplicity of design and performance characteristics. Microstrip antennas have authentic applications in several fields like the medical field, military systems such as aircraft and spacecraft, missiles, remote sensing, and rockets. Since it has a low profile, it is also used in satellite communication. Nowadays, from a commercial perspective, the demand for microstrip antennas has increased due to its cheap substrate material and easy fabrication process. Some other applications of microstrip antennas are discussed below [24]:

(a) Mobile and Satellite Communication Application: Microstrip antenna has small size, low profile, and low cost due to which it is commonly used for mobile communication system application. In satellite communication, we generally require circularly polarized radiation patterns for the operation, which can be easily achieved by MSA [25].

(b) Global Positioning System Applications: In the current era, microstrip patch antennas are also used for global positioning systems (GPS) with the help of high permittivity sintered substrate material. The two important main features are circular polarization and compactness that make it capable of GPS applications.

(c) Radio Frequency Identification (RFID): The RFID system is a transponder or tag and a reader or transceiver. It can be used in different fields such as mobile communication, manufacturing, logistics, transportation, car parking, animal tracking, and health care. The RFID system works between 30Hz to 5.8GHz frequency.

(d) Interoperability for Microwave Access (WiMax): Another name of IEEE 802.16 standard is WiMax. An attractive feature of WiMax is that it easily provides a data rate of up to 70 Mbps and coverage up to the radius of 30 miles. Microstrip patch antenna is used in WiMax because it produces three important resonant modes at 2.7 GHz, 3.3 GHz, and 5.3 GHz.

(e) RADAR Application: Radar is commonly used for the detection of stationary/static objects and dynamic/non-stationary objects like vehicles, people, spacecraft and aircraft, *etc.* Photolithography is used for the fabrication of microstrip antenna. That is why the production at a large scale is possible along with repeated accuracy and precision at a cheaper cost and in lesser time as compared to other antennas. Therefore, a microstrip antenna is an ideal and attractive choice for radar applications.

(f) Bluetooth Applications: Microstrip antenna is perfectly suitable for Bluetooth applications due to its reduced size. It works in the range of 2400 to 2484 MHz (called ISM Band). In the current era, an aired substrate is used, as a result of which this type of antenna occupies a very less volume of $33.3 \times 6.6 \times 0.8$ mm^3, making it suitable for Bluetooth application.

(g) Wireless Communication Applications: The microstrip patch antenna having an S-shaped patch is frequently used in wireless communication systems due to suitable quality parameters. This S-shaped patch antenna is formed by inserting two slots into a rotated square patch, after which it looks like an English letter 'S', which is why it is called an S-shaped patch antenna. The enhancement of bandwidth is due to the presence of slots and thick substrate; therefore microstrip antenna is an attractive and ideal choice for broadband [26, 27].

CONCLUSION

In this chapter, fundamental of microstrip antenna has been discussed with the help of geometry. A comparative study of various feeding techniques has been done along with their advantages and disadvantages. In this chapter designing a microstrip antenna is also discussed in a very lucid manner. The microstrip patch antennas are very attractive due to their simplicity of design procedure. Microstrip antennas have various important applications in several fields like the medical field, military systems such as aircraft and spacecraft, satellites, missiles, remote sensing, and rockets.

CONSENT FOR PUBLICATION

Not Applicable.

CONFLICT OF INTEREST

The author confirms that this chapter contents have no conflict of interest.

ACKNOWLEDGEMENT

Declared none.

REFERENCES

[1]　A. Constantine, *Antenna Theory Analysis and Design* Wiley & Sons, INC., Publications..

[2]　R.E. Munson, "Conformal Microstrip Antennas and Microstrip Phased Arrays", *IEEE Trans. Antenn. Propag.,* vol. AP-22, pp. 74-78, 1974.
[http://dx.doi.org/10.1109/TAP.1974.1140723]

[3]　N. Mahmoud, and E.K.I. Hamad, "Tri-band microstrip antenna with L-shaped slots for Bluetooth/WLAN/WiMAX applications", *33rd National Radio Science Conference (NRSC),* 2016pp. 73-80
[http://dx.doi.org/10.1109/NRSC.2016.7450826]

[4]　D. Kranti, "Patil, Design of Advanced Microstrip Antenna for Wearable Application", *Journal of Signal Processing,* vol. 6, no. 1, pp. 1-4, 2020.

[5]　K.K. Naik, and A.V.S. Pasumarthi, "Design of Hexadecagon Circular Patch Antenna with DGS at Ku band for Satellite communications", *Progress In Electromagnetics Research M,* vol. 63, pp. 163-173, 2018.
[http://dx.doi.org/10.2528/PIERM17092205]

[6]　Shaktijeet Mahapatra, and Mihir Mohanty, *Design of Novel Multi-band Rectangular Patch Antenna for Wireless Communications.,* 2020.
[http://dx.doi.org/10.1007/978-981-13-8461-5_4]

[7]　L. Dang, Z.Y. Lei, Y.J. Xie, G.L. Ning, and J. Fan, "A Compact Microstrip Slot Triple-Band Antenna for WLAN/WiMAX Applications", *IEEE Antennas Wirel. Propag. Lett.,* vol. 9, pp. 1178-1181, 2010.
[http://dx.doi.org/10.1109/LAWP.2010.2098433]

[8]　Z. Zhou, Z. Wei, Z. Tang, and Y. Yin, "Design and Analysis of a Wideband Multiple-Microstrip Dipole Antenna with High Isolation", *IEEE Antennas Wirel. Propag. Lett.,* vol. 18, no. 4, pp. 722-726, 2019.
[http://dx.doi.org/10.1109/LAWP.2019.2901838]

[9]　Q. Chen, H. Zhang, X. Zhang, M. Jin, and W. Wang, "An X Band Dual-polarized Shared Aperture Antenna Using Waveguide and Microstrip Antennas", *2018 IEEE Asia-Pacific Conference on Antennas and Propagation (APCAP), Auckland,* pp. 106-107, 2018.
[http://dx.doi.org/10.1109/APCAP.2018.8538018]

[10]　S. Rashid, and C.K. Chakrabarty, "Bandwidth Enhanced Rectangular Patch Antenna Using Partial Ground Plane Method for WLAN Applications", *Universiti Tenaga Nasional, Putrajaya Campus. The 3rd National Graduate Conference,* 2015.

[11]　R. Garg, P. Bhartia, and I. Bahl, *Ittipiboon, "A., Microstrip Antenna Design Handbook.* Artech House, Inc, 2001.

[12]　F. Yang, X. Zhang, X. Ye, and Y. Rahmat-Samii, "Wide-band E-shaped patch antennas for wireless communications", *IEEE Trans. Antenn. Propag.,* vol. 49, no. 7, pp. 1094-1100, 2001.
[http://dx.doi.org/10.1109/8.933489]

[13]　Y. Ge, K. Esselle, and T. Bird, "E-shaped patch antennas for high speed wireless networks", *IEEE Trans. Antenn. Propag.,* vol. 52, no. 12, pp. 3213-3219, 2004.
[http://dx.doi.org/10.1109/TAP.2004.836412]

[14]　S.K. Behera, "Novel Tuned Rectangular Patch Antenna as a Load for Phase Power Combining", *PhD Thesis, Jadavpur University, Kolkata,* 2012.

[15]　G. Singh, and J. Singh, " Comparative Analysis of Microstrip Patch Antenna with Different Feeding Techniques", *International Conference on Recent Advances and Future Trends in Information Technology, Proceedings published in International Journal of Computer Applications® (IJCA),* 2012.

[16]　K. Dipak, "Neog, Shyam S. Pattnaik, Dhruba. C. Panda, Swapna Devi, BonomaliKhuntia, and Malaya

Dutta, Design of a Wideband Microstrip Antenna and the Use of Artificial Neural Networks in Parameter Calculation", *IEEE Antennas Propag. Mag.,* vol. 47, no. 3, 2005.

[17] G. Singh, and J. Singh, "Comparative Analysis of Microstrip Patch Antenna with Different Feeding Techniques", *International Conference on Recent Advances and Future Trends in Information Technology,* pp. 18-22, 2012.

[18] F.S. Fong, H.F. Pues, and M.J. Withers, "Wideband multilayer coaxial-fed microstrip antenna element", *Electron. Lett,* vol. 21, pp. 497-498, 1985.
[http://dx.doi.org/10.1049/el:19850352]

[19] Y. Gupta, "Stacked Microstrip Patch Antenna with Defected Ground Structures for W-lan and Wimax Applications", *Thesis work in Department of Electronics and Communication Engineering, Thapar University, Patiala, June,* 2014.

[20] C. Rai, A. Raghuwanshi, and S. Lal, "A Review on Microstrip Patch Antenna", *Int. J. Res. Appl. Sci. Eng. Technol.,* vol. 5, no. 5, pp. 1279-1281, 2017.

[21] Sandip Gajera, Hasani Mahesh, and Naresh Patel, "Comparative Analysis of Microstrip Line and Coaxial Feeding Technique for Rectangular Microstrip Patch Antenna", *International Journal of Electronics and Computer Science Engineering,* vol. 2, no. 2, pp. 719-725, .

[22] V. Thakur, and S. Kashyap, "A Review Paper on Techniques and Design for Microstrip Patch Antenna", *International Journal of Advanced Research in Electrical, Electronics and Instrumentation Engineering,* vol. 4, no. 2, pp. 656-662, 2015.
[http://dx.doi.org/10.15662/ijareeie.2015.0402022]

[23] "Ashwin khandelwal, Anushka Bhurke and Rahul Koshti, "A Review on Optimization of microstrip Patch Antenna," International Journal of Innovative Research in Science", *Engineering and Technology,* vol. 6, no. 11, pp. 21243-21251, 2017.

[24] B.D. Patel, "Tanisha Narang, Microstrip Patch Antenna- A Historical Perspective of the development", *Conference on Advances in Communication and Control Systems,* 2013pp. 445-449

[25] P.K. Malik, D.S. Wadhwa, and J.S. Khinda, "A Survey of Device to Device and Cooperative Communication for the Future Cellular Networks", *Int. J. Wirel. Inf. Netw.,* 2020.
[http://dx.doi.org/10.1007/s10776-020-00482-8]

[26] Kr. Praveen, "Hardware Design of Equilateral Triangular Microstrip Antenna Using Artificial Neural Network", *TECHNIA – International Journal of Computing Science and Communication Technologies,* vol. 3, no. 2, pp. 0974-3375, 2011.

[27] Tripathi M.P, Malik P, and Parthasarthy H, "Axisymmetric Excited Integral Equation Using Moment Method for Plane Circular disk", *International Journal of Scientific and Engineering Research,* vol. 3, no. 3, pp. 1-3, 2012.

CHAPTER 2

Fractal and Defected Ground Structure Microstrip Antenna for Wireless Applications

Charu Tyagi[1,*], **Puneet Chandra Srivastava**[1], **Patrika Jayanti**[2] and **Kiran Srivastava**[3]

[1] Department of Electronics and Communication Engg., Raj Kumar Goel Institute of Technology, Ghaziabad, India

[2] Department of Electronics Engg., Mahamaya Polytechnic of IT, Hathras, Uttar Pradesh 202002, India

[3] Department of Electrical and Electronics Engg., Galgotias College of Engineering and Technology, Greater Noida, India

Abstract: The new technical developments in wireless networking devices require simultaneous operation at different frequencies for different applications such as LTE, IoT, wi-fi, *etc.* To achieve the compact microchip scale, this chapter explores the numerous key features of miniaturization techniques in basic antenna with advanced fractal arts. Fractals are a subset of non-Euclidian geometry. The larger bandwidth with multi frequencies is a key characteristic of antennas that have a fractal design in them. The defective ground design successfully overcomes the demerits associated with fractal design in the antenna. Modern-day applications like IEEE802.16 (Wi-Max, WLAN), Mobile operating bands (GSM, LTE), BT, *etc.*, are covered in this chapter.

Keywords: Defected ground, Fractal geometry, Multiple inputs, Multiple outputs, Wireless communication.

1. INTRODUCTION

The latest rising demand in the defense and telecommunication industry is antenna miniaturization. Year after year, with the developments in miniaturization technology and market demand, antenna design has also been optimized with numerous new techniques [1]. Fractal is one of the common techniques with relevance in the new era of system or equipment miniaturization. Mandelbrot's

---

* **Corresponding author Charu Tyagi1:** Department of Electronics and Communication Engg., Raj Kumar Goel Institute of Technology, Ghaziabad. E-mail: charutyagi24@gmail.com

Praveen Kumar Malik, Pradeep Kumar, Sachin Kumar and Dushyant Kumar Singh (Eds.)

first toggled fractal work defined the broken or irregular fragments [2, 3]. Mandelbrot identified a series of complex shapes in their Euclidean geometrical structures that are inherently similar. The fractal has now successfully proven its influence on the broad horizon of examining natural objects such as clusters, mountain ranges, trees, cloud borders, coastlines, *etc.* With the latest fractal model techniques, the issue of physical size reduction was solved, but the new problems began with electromagnetic theory that need to be explored to make antenna designs effectively miniaturized. In miniaturizing the size of the antenna, some of these geometries were particularly helpful, while other designs targeted multi-band parameters [4, 5]. These fractal antennas are low- profile with modest gain and are designed to work at a wide range of frequency bands such that they can be used for multi-functional device structures. The fractal has now successfully demonstrated its influence on the broad horizon for analyzing natural things such as clusters, mountain ranges, trees, cloud borders, coastlines, *etc* [6, 7]. The topic of physical size reduction has been discussed in the chapter as a literary exploration of fractal theory.

Evolution in Antenna Optimization

The term "fractal" is taken from the Latin word 'fractus' that was first used by Mandelbrot in the early 1970s for 'broken', *i.e.*, fractional, irregular, or fractured. Fractals are normally self-similar, independent of size. Therefore, some similar characteristics of fractal geometric shapes are:

- Self-similarity of original geometry sub-parts
- The new dimension of the main geometry is fractional.

These fractal geometries are used to describe natural structures that, with the aid of normal Euclidean geometries, are very difficult to define. The density of clouds, the branching of trees, the length of a coastline, and many more are examples, just as nature does not restrict itself to Euclidean geometries [8, 9].

2. FRACTAL LITERATURE

Fractal geometries are a distinctive class of shapes that have never been described based on standard sizes. All fractal geometry consists of a single elementary fractal form with many iterations. The iterations continue to be infinite, creating a form with infinite dimensions within a finite boundary. In lightweight telecommunication devices used for wireless communication applications, this miniaturizing property is of great importance since a smaller receiver helps to reduce the total device size.

3. USEFUL FRACTAL GEOMETRIES

Due to their complex erratic behavior, fractal geometries were originally discarded, but later Mandelbrot discovered certain special properties associated with them which makes these geometries useful to model problems in the real world. Maximum fractal curves were well defined earlier and were associated with later-year mathematicians, but Mandelbrot's research outcome was path-breaking. The model complex geometries by finding the same feature in several irregular geometries looked similar are presented in literature based on his findings. The Sierpinski fractal gasket was the first fractal that is shown in Fig. (**1**) as common, the design procedure for forming this fractal starts with an equilateral triangle positioned in the plane as shown in Fig. (**1**) stage 0. The further steps in the design process (see phase 1 of Fig. **1**) delete the central triangle region with vertices contained in the center points of the sides of the original triangle, as seen in phase 0.

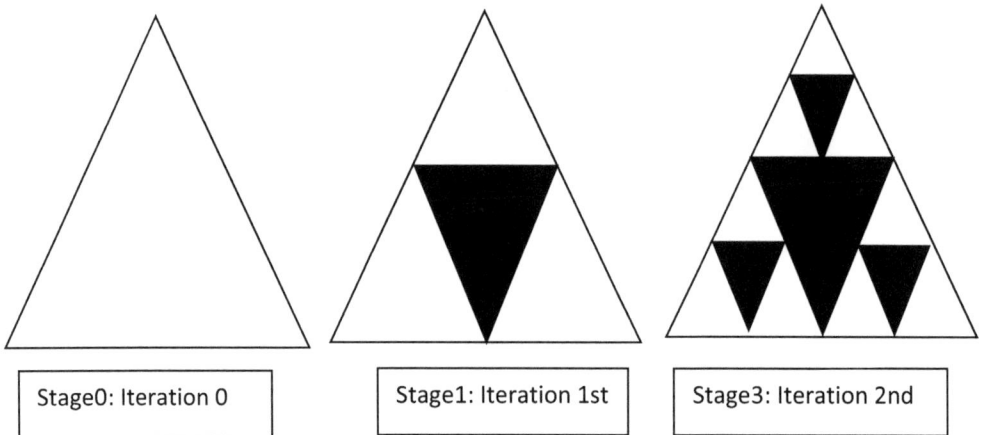

| Stage0: Iteration 0 | Stage1: Iteration 1st | Stage3: Iteration 2nd |

Fig. (1). Multiple stages in the sierpinski gasket fractal formation.

As in the 2nd stage of Fig. (**1**), the design process is continued repeatedly until the targeted design has been accomplished. Through this iterative process, the Sierpinski fractal gasket can be created at infinite times to miniaturize the framework. This description explicitly shows that the Sierpinski gasket is a form of fractal that is self-like. A conducting metal is a white triangular area in Fig. (**1**), while the dark triangular areas reflect regions where metal has been stripped.

Another fractal widely used in antenna architecture is the Koch snowflake fractal. In the starting plane, as followed in stage 0 of Fig. (**2**), this fractal also has a good equilateral triangle basis.

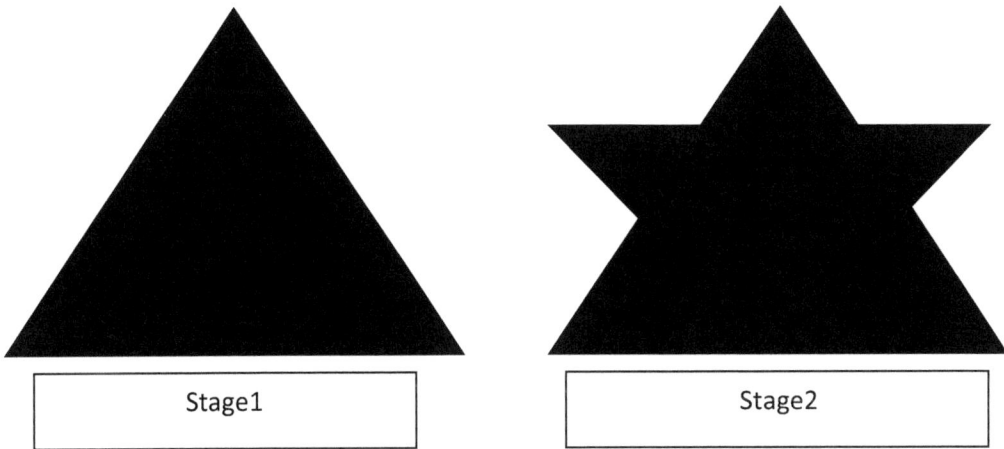

Fig. (2). Initial iterative stages of koch snowflake fractal design.

The Koch Snowflake Fractal is created by attaching smaller and smaller triangles to the structure in an iterative way in a given sequence of dimensions, far from the Sierpinski gasket fractal, which was built by eliminating small systematic geometry with the base configuration. In Fig. (**2**), where the initial phases are shown for a clear understanding of the Koch snowflake design, the design procedure is notably demonstrated.

In the development of new designs for antennas, fractal trees have proved to be very useful in random geometry along with other numbers of fractal forms. The fractal structure shown in Fig. (**3**) is particularly composed of tree branches that are similar to the fractal configuration of the Sierpinski gasket.

The electronic bandgap in parameter optimization of the antenna is a very common technique. With the support of space, the EBG structure successfully achieved the space-filling properties of fractal geometry. Antenna models are also fascinated by the Hilbert curve or related curves. The iterations of the building of the Hilbert curve are illustrated in Fig. (**4**).

The Koch fractal is another widely used fractal to optimize the ground and radiation pattern of the antenna. Fig. (**5**) displays few other Koch curves, snowflakes, Koch islands, *etc.* which are most common in the miniaturization of antenna architecture. Fractal trees and Koch curves are also used for dipole antenna miniaturization. In the configuration of a low-profile multi-band antenna, fractal space-filled properties are very useful.

Fig. (3). Fractal Trees.

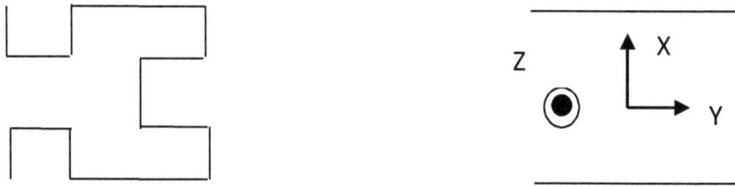

Fig. (4). Hilbert Curve (curve construction).

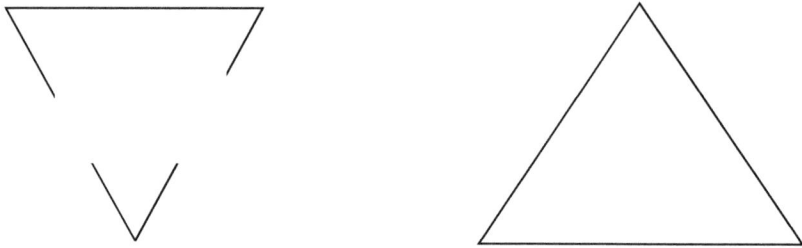

Fig. (5). Koch Snowflake.

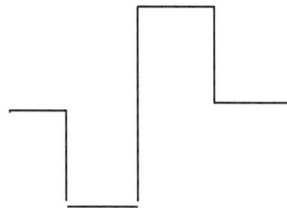

Fig. (6). Minkowski Curve Fractal.

4. MINKOWSKI CURVE FRACTAL

Minkowski sausage is more generally known as the Minkowski Curve. Mandelbrot claimed that the root of the curve is not known, and Hermann

Minkowski developed it. Like fractal curves, the construction of this curve is also based on a repeated process. The simple curve begins with a straight line, for each line segment of the eight-sided curve generator. There are 8 separate segments.

Fig. (7). 1st iteration of Minkowski curve.

The same generator of 8 segments will be added to create slightly more complex curves at each branch of the base curve as shown in Figs. (**6** and **7**).

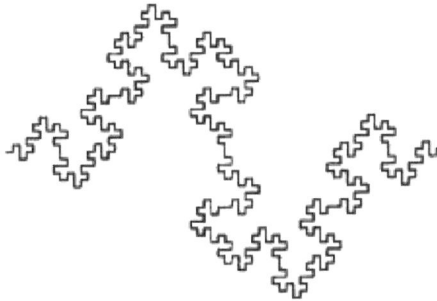

Fig. (8). 2nd Iteration of Minkowski curve.

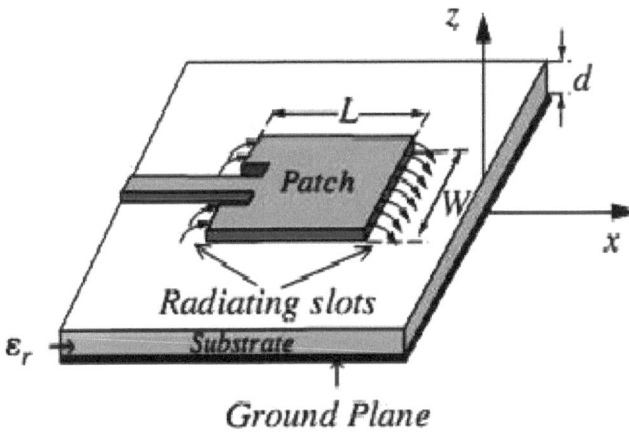

Fig. (9). 3rd Iteration of Minkowski curve.

Each iteration gives the next miniaturization step. After the first iteration of the modeling process, the appearance of the curve is changed considerably as shown in Fig. (**8**).

4.1. Properties

4.1.1. Curve Length

Each repetition extends the length of the curve. The segments are multiplied by eight for every iteration and the length segments are separated by four. The length of the curve reaches infinity with expanded iterations.

4.1.2. Fractal Dimension

- The fractal dimension is computed using Hausdorff-Besicovitch
- Dimension = log (Multiply factor segment 'X') / log (divider segment length 'D')
- In the above equation, D is replaced by 4 (since each segment at the next iteration level is divided by 4) & X by 8 (since the drawing process gives eight segments). Therefore, in the Hausdorff-Besicovitch equation, it gives:
- Dimension for iteration = log(8) / log(4) = 1.5

4.1.3. Self-Similarity

We found that every iteration generates a similar shape with a smaller dimension. The increasing iteration labels increase the electrical length of the antenna and every small segment is self-similar with other same small segments.

4.2. How Fractal have Space Occupancy Architecture

Regular Euclidean geometry is restricted to points, surfaces, sheets, and volumes, while somewhere among them lies fractal structures. It can be concluded that a fractal is a line, tending to a sheet with an increased iteration level. The novel property of occupancy of space creates a curve that is electrically very long but fits in a narrower physical space.

Space-filling characteristics improve the antenna's radiation length, which is very useful in design miniaturization. The fractal curve of Minkowski has a similar property and for each repetition, the electrical length is expanded by one-third of its real length. Minkowski Curve's N^{th} Order Iterations can be formulated as:

Duration N = (4/3)N * Original Duration (Lo)

After N^{th} Iteration & Lo, LN is the electrical length before any iteration of the original length. The circumferential region of solid fragment and defected area modified with an improvement in iteration order in Minkowski fractal.

5. ADVANTAGES AND DISADVANTAGES

Fractal antennas, enumerated below, have various proven advantages:

- Reduced cross-sectional area
- Non-requirement of Network matching Impedance
- Multi resonance frequencies
- Optimized Gain & directivity

There are two significant drawbacks of these antenna designs in the early production stages:

- Development and manufacturing have found little complexity.
- Decreases in antenna gain with increased iteration

The earlier drawbacks of fractal architectures have been solved by more continuous study and use of hybrid technology. With EBG architectures (Defected Ground), Parasitic patches, array designs, and several more strategies coupled with foundation designs, lower gain and directivity are largely optimized.

6. NATURAL FRACTAL AND ITS APPLICATIONS

Fractals are not limited to pretty images and complicated computer-generated shapes. Any irregular and random look is referred to as a fractal shape. Fractals appear as small as the cell nucleus and as large as the solar system in our daily lives. In theory, everything that exists in the world can be described as a fractal, with extensive applications in science.

6.1. Sciences of Nature

Fractals are an explanation similar to that of the universe. As per the assumption of a cosmologist, the matter is uniformly spread across space. The observation contradicts the assumption stated above. Nevertheless, many scientists claim that the structure of the universe on all scales is more or less fractal.

6.2. Of Nature

Select a tree, for example. For the closer study, select a particular branch. From that branch, pick a bunch of leaves. The tree, the branch, and the leaves are all identical objects mentioned. Unpredictability, randomness, and messiness are suggested in this example. The cloud weather example is quoted by some individuals. The majority of things present in nature can be visually modeled by fractal geometries, the most recognized being mountains, coastlines, and clouds. Fractals are also useful for modeling soil erosion and simultaneously analyzing seismic patterns.

6.3. Computer-Science

Image compression using fractals in computer science is the most potent application of fractals. The facts that the real world is well explained are used in this advanced compression. Compression of images is much more than traditional (*e.g.* .mp4, .jpg, .gif, *etc.*) compression by using fractal.

6.4. Fluid- Mechanics

The turbulence of the flow is more adapted to fractals. Turbulent flows are messy and fairly difficult to model in the study of fluid mechanics. For scientists to better understand the model complex flows, the fractal theory is helpful. Flow simulation is also possible through the theory of fractal geometries and is widely used in petroleum science too.

7. TELECOMMUNICATIONS

The size of devices used in telecommunications has been reduced with the evolution of the fabricated antenna. This fractal geometry can be broadly covered by the new requirement. Fractal attraction is dependent on multi-frequency, smaller size with high directivity, and so on. The fractal part generally provides 'fractal loading' and generates the smaller antenna for a chosen frequency of use.

Without efficiency, a realistic size reduction of 2-4 times is feasible. With the help of a few extra methods, surprisingly high performance is achieved.

7.1. Introduction to Defected Ground Structures

According to its ease of manufacturing, compact scale, linear & circular polarization, roughness when placed on rigid surfaces, the microstrip antenna has various uses relative to other types of antennas. Due to narrow bandwidth, lower performance, surface wave loss and lower gain, it has its own limitations along with merits. The defective material ground produces an electrical band differences that overcomes the limitations of the traditional patch antenna supersede as shown in Fig. (**6**).

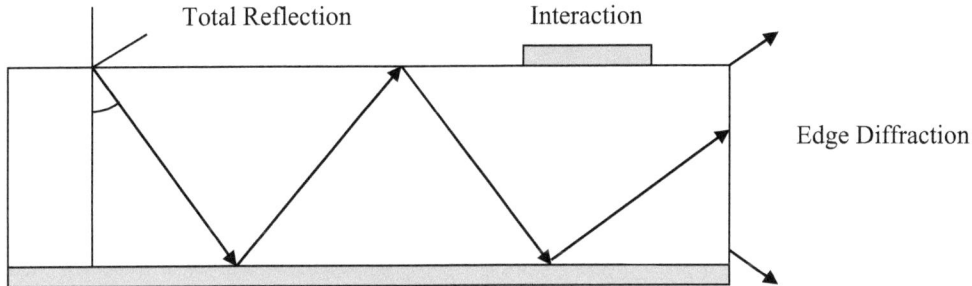

Fig. (10). Microstrip antenna with finite ground.

Alternating currents of a particular frequency spectrum cannot spread through deficient ground planes such as regular conducting planes, distinguished by high impedance surfaces that do not support the transmission of surface waves through them as well. The latest ground designs have demonstrated high benefit in output enhancement or features such as radiation pattern broad bandwidth along with reduced coupling effect in close weapons. The reflection phase in the defective ground structure varies continuously from -180 ° to 180 ° *versus* frequency, which makes it relevant for the low profile wire antenna design. Earlier EBG structures have physical size limitations, particularly at the low operating frequency, but it is more important and functional with the advent of fractal geometries.

7.2. The Ground Plane Impact

The ground plane should be the eternal property of the antenna in ideal condition, though it is not feasible on practical grounds. The patch antenna should be smaller since it should not stretch its ground plane past the boundaries. The radiation that is produced between the ground and the patch is connected to the antenna fringing area. Miniaturization of the antenna size is based on the thickness of the dielectric substrate. The ground of the patch antenna should be greater by $\lambda/4$ from its sides, while the total radius of the ground plane should be at least λ. The ground plane is the main component of the antenna and in an optimal situation, it can be an infinite mono-pole antenna. In practice, a smaller ground plane is desirable. The radiation in the patch is created by the fringing field between the patch and the ground plane. Conventional patch antennas presume exponentially large dimensions of the ground plane, creating greater antenna sizes that are inefficient in smaller modern designs. In radiation patterns, bigger ground planes establish higher directivity. Because the antenna patch needs to have finite ground in practical terms, its effect should be optimized to achieve the desired results. The GTD (Uniform Geometrical Diffraction Theory) was used from the antenna patch to measure the edge diffracted fields forming the finite ground plane. To measure the source field of the radiating patch, two separate methods were used in the theory of modal expansion and the theory of slot. Many analytical test results were presented to demonstrate the effect of the finite ground plane [5].

The problem with planar antennas was that plane waves came from a plane interface between two distinct media: dielectrics-dielectrics and conductor-dielectrics. Researchers have worked on various innovative approaches over the past decades to maximize the reduction of power transfer during patch excitation by surface waves as shown in Fig. (**11**). There are two limitations to the planner antenna, one is conductor dielectrics, and the other is dielectric-dielectric. Theoretically, the waves spread across the above-mentioned interfaces, the planner antenna's main problems are the optimization of propagation over the boundaries. The disturbances over the borders do not move the surface wave power into the patch's primary radiation, the default power dispersed over the ground plane allows radiation patterns to repulse and nulls, a lower gain of low polarization, and increase in the back radiation [10].

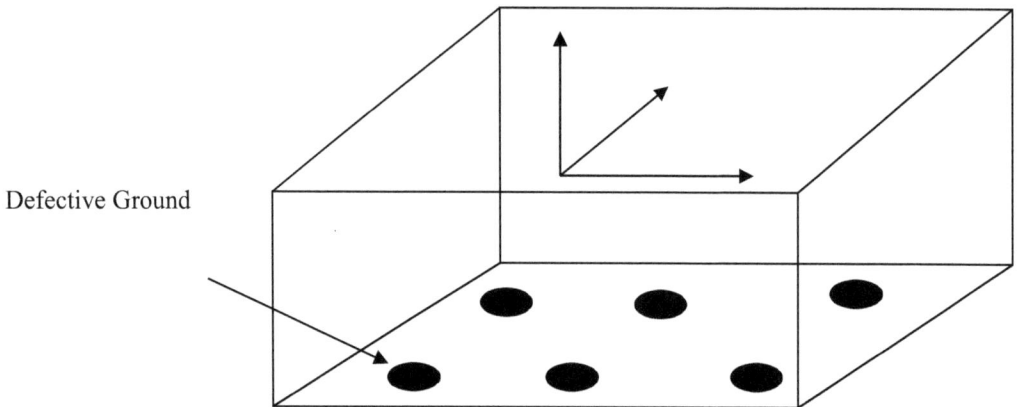

Defective Ground

Fig. (11). Surface waves propagation in patch antenna.

In antenna patches, deficient ground structures allow for intense suppression of surface waves. EBG has developed a promising approach to solve the problems (low directionality / low gain) and to make it practical to build lightweight antenna arrays with high gain with the desired radiation properties. The proper installation of Defected Land Systems is the secret to progress.

In certain respects, defective field structures may be used. Periodic substrate loading is one of the techniques widely used in a particular frequency range to produce surface wave dispersion. The basic configuration of the deleted ground system increases the antenna's radiating power that couples with the current space wave because the substratum does not allow the propagation of the surface wave across it. Defected ground resonator antennas are the second common method, which is technically very efficient in obtaining patch antenna enhancement. With the insertion of a deleted metallic patch positioned over the other antenna in resonator antennas to form a resonant cavity that raises the small antenna parameters (Fig. **11**). A variety of experiments are being conducted around the world to miniaturize antenna size for recent wireless communication devices such as smartphone, desktop, IoT, RADAR, *etc.* along with multiband, super-wide array, high gain high directivity. The communication equipment used by cellular, WLAN, GSM, 4G, 5G, and Bluetooth includes smaller multiband antenna sizes. To better explain miniaturization, multiband and ultra-large band characteristics using various fractal geometries with defected ground architecture, few examples have been carried out *via* this article.

7.3. Multiband Antenna using Koch and Sierpinski Fractal with Defected Ground

We needed the multifunctional system with minimal space acquired, as per current technical trends. Antenna miniaturization is the secret to the multi-frequency operation of today's systems. With hybrid fractal geometry, which incorporates two fractal geometries to obtain optimized antenna characteristics, the miniaturization of the antenna may be accomplished [11].

The antenna discussed here has the composite radiator of a second stage iteration of the Koch snowflake with a fourth phase Sierpinski Carpet fractal slotted iteration within it. The deleted ground approach used for the six cornered ring provides the six resonant frequencies. The antenna developed on the FR4 substrate is 80 mm x 54 mm x 1.6 mm in height. The substrate used has a tangent loss of 0.02. For multiple multiband frequencies from 850 MHz to 5.99 GHz, the calculated outcome illustrated omnidirectional radiation patterns with gains ranging from 2.9 to 4.64 dBi.

7.4. Ultra-Wideband Antenna using Fractal

Multimedia development requires a higher data transmission rate, which can only be supported by ultra-wideband antennas. It is possible to obtain the ultra-wideband using the hybrid fractal form. This article explored using a particular geometry of the 4th iteration stage of the flower form. With a peak gain of 5dB, the design reached a 3db bandwidth of 12.12 GHz (frequency ranged from 4.58-16.7 GHz). Antenna with tangent loss 0.02 for patch scale 15.7mm x 11.4mm x 1.6mm using the FR4 substrate content [12].

7.5. S & C Band Antenna with Defective Ground

Wireless sensor networks include handheld wireless applications, with a need for smaller microchip scale antennas that are lighter in weight for wireless communication applications such as mobile ad-hoc networks (MANET), Delay Tolerant Networks, (DTN). The discussed antenna geometry is designed on FR4 epoxy substrate with loss tangent 0.002 the antenna used line feeding with 5 mm lumped ports. The antenna parameters were optimized by rendering the ground defective. The calculated result is 2.83 GHz with a peak gain of 6.83dB. For devices running in the S&C band, the 3dB bandwidth ranging from 2.05 to 4.88 GHz is sufficient [13].

7.6. Miniaturization of Ultra-wideband Antenna with Fractal Iteration

Higher spectral bandwidth and higher data transmission rates are posing significant challenges to traditional antennas, which has shifted the focus to miniaturised ultra-wideband (UWB) antennas. The antenna miniaturization is achieved by deleting Sierpinski square slots in a radiating patch in an existing decagonal-shaped antenna as shown in Table **1**. Compared to 123 percent of the base architecture, the Fractal optimization allows reaching 127 percent bandwidth (frequency varying from 15.1 GHz). The volume reduces to approximately 36% and the area reduces to 46.3% [**14**].

Table 1. Characteristics of different fractal antenna.

S.no	Substrate Material	Size of Antenna (LXWXH mm³)	Bandwidth	Frequency Band	Gain (dB)	Designing Technology
1	FR-4 (h=1.6mm, εr = 4.4, tan δ = 0.02)	80x54 x1.6	Multiband antenna	0.85–0.98 GHz	1.4-4.64	Hybrid fractal with Defected Ground
		50.8x27× 1.6		1.27–1.52 GHz		
		2x47× 1.6		1.82–2.08 GHz		
		4.6x2 × 1.6		2.40–3.05 GHz		
		18x 54× 1.6		3.64–3.92 GHz		
				4.74–5.99 GHz		

(Table 1) cont.....

S.no	Substrate Material	Size of Antenna (LXWXH mm³)	Bandwidth	Frequency Band	Gain (dB)	Designing Technology
2	FR-4 (h=1.6mm,εr = 4.4, tan δ = 0.02)	15.7 × 11.4 × 1.6	Ultra wide band	4.58-16.7 GHz	1.3-4.8	Waveguide feeding with Fractal antenna
3	FR-4 (h=1.6mm,εr = 4.4, tan δ = 0.02)	32 × 30 × 1.6	Wideband	2.05-4.80GHz 2830MH-BW	1.2-7.4	Defected Fractal Ground
4	FR-4 (h=1.6mm,εr = 4.4, tan δ = 0.002)	35 x 35 x 1.6	Ultra wideband	3.5-15.1GHz	1.4-5.9	Miniaturization with Fractal iterations & Deleted Ground

During the implementation of irregular geometries in basic antenna designs, attention was paid in two areas. The first area explores design analysis of fractal elements [15, 16] concerns and the second area is exploring the application in array designing. The recursive nature of FA leads to the development of rapid beamforming as per algorithms. Because of ultra-wideband characteristics, the most of FA is used in imaging systems and compact automobile radar systems [17, 18]. The recent advancement in FA geometries finds application in upcoming communication bands and will strongly impact the 5th generation communication equipment [19].

No significant impact was observed in the early development of different Fractal models to meet the various applications in terms of bandwidth and gain, and would need further optimization to achieve better performances [20]. For the targeted frequency band of 3.1 GHz to 10.6 GHz, the antenna discussed found improved Impedance matching with constant gain [21]. This design was also validated for Current Distribution, group delay, radiation pattern and found satisfactory results.

The exploration of future advancement can be done based on two suggested points, one is the concept of arrays, and the other is the layered architecture of fractal [22]. Layered fractal can be achieved by a two-layer fractal [22], so the original size will be intact. The antenna element expansion also adds a significant improvement in the performance of array antennas. These designs can be successfully mounted for on-body applications [23], automobile devices, *etc.*

CONCLUSION

In recent years, the wireless communication industry has grown rapidly, attracting researchers in the area of antenna miniaturization The defected ground antenna gain optimised with multiband frequency using fractal and defected ground can be used to create a compact chip size multiband antenna architecture.

CONSENT FOR PUBLICATION

Not Applicable.

CONFLICT OF INTEREST

The author confirms that this chapter contents have no conflict of interest.

ACKNOWLEDGEMENT

Declared none.

REFERENCES

[1] M. Alibakhshikenari, M. Khalily, B.S. Virdee, C.H. See, R.A. Abd-Alhameed, and E. Limiti, "Mutual Coupling Suppression Between Two Closely Placed Microstrip Patches Using EM-Bandgap Metamaterial Fractal Loading", *IEEE Access,* vol. 7, pp. 23606-23614, 2019. [http://dx.doi.org/10.1109/ACCESS.2019.2899326]

[2] A.T. Abed, and A.M. Jawad, "Compact Size MIMO Amer Fractal Slot Antenna for 3G, LTE (4G), WLAN, WiMAX, ISM and 5G Communications", *IEEE Access,* vol. 7, pp. 125542-125551, 2019. [http://dx.doi.org/10.1109/ACCESS.2019.2938802]

[3] H. Ullah, and F.A. Tahir, "A Novel Snowflake Fractal Antenna for Dual-Beam Applications in 28 GHz Band", *IEEE Access,* vol. 8, pp. 19873-19879, 2020. [http://dx.doi.org/10.1109/ACCESS.2020.2968619]

[4] M. Naghshvarian-Jahromi, "Novel Wideband Planar Fractal Monopole Antenna", *IEEE Trans. Antenn. Propag.,* vol. 56, no. 12, pp. 3844-3849, 2008. [http://dx.doi.org/10.1109/TAP.2008.2007393]

[5] W. Balani, "Design Techniques of Super-Wideband Antenna–Existing and Future Prospective", *IEEE Access,* vol. 7, pp. 141241-141257, 2019. [http://dx.doi.org/10.1109/ACCESS.2019.2943655]

[6] W. Cheng, J. Wang, R. Zhang, and H. Zhang, "Achieving Maximum Spectrum Efficiency for Full-Duplex MIMO: The Fractal Array-Element Based Self-Interference Mitigation Approach", *IEEE Access,* vol. 7, pp. 74056-74069, 2019. [http://dx.doi.org/10.1109/ACCESS.2019.2920431]

[7] S. Kumar, A.S. Dixit, R.R. Malekar, H.D. Raut, and L.K. Shevada, "Fifth Generation Antennas: A Comprehensive Review of Design and Performance Enhancement Techniques", *IEEE Access,* vol. 8, pp. 163568-163593, 2020. [http://dx.doi.org/10.1109/ACCESS.2020.3020952]

[8] J.B. Benavides, R.A. Lituma, P.A. Chasi, and L.F. Guerrero, "A Novel Modified Hexagonal Shaped Fractal Antenna with Multi Band Notch Characteristics for UWB Applications", *IEEE-APS Topical Conference on Antennas and Propagation in Wireless Communications (APWC),* 2018

[http://dx.doi.org/10.1109/APWC.2018.8503774]

[9] Abhik Gorai, Manimala Pal, and Rowdra Ghatak, "A Compact Fractal-Shaped Antenna for Ultrawideband and Bluetooth Wireless Systems With WLAN Rejection Functionality", *IEEE Antennas and Wireless Propagation Letters,* vol. 16, pp. 2163-2166, 2017.
[http://dx.doi.org/10.1109/LAWP.2017.2702208]

[10] K. Johnson, "Digital Manufacturing of Pathologically-Complex 3D Printed Antennas", *IEEE Access,* vol. 7, pp. 39378-39389, 2019.
[http://dx.doi.org/10.1109/ACCESS.2019.2906868]

[11] Yu Z., Yu J., Ran X., and Zhu C., "A novel Koch and Sierpinski combined fractal antenna for 2G/3G/4G/5G/WLAN/navigation applications", *Microwave and Optical Technology Letters,* vol. 59, pp. 2147-2155, 2017.

[12] S. Singhal, and A.K. Singh, "Flower Shaped Ultra Wideband Fractal Antenna", *International Journal of RF and Microwave Computer-Aided Engineering,* pp. 1-8, 2019. Wiley.

[13] Swati Jindal, and Jagtar Singh, "Defected Ground Based Fractal Antenna for S and C Band Applications", *Wireless Pers Commun,* vol. 110, pp. 109-124, 2020.

[14] "Tanweer Ali*, Subhash B K, and Rajashekhar C. Biradar "A Miniaturized Decagonal Sierpinski UWB Fractal Antenna"", *Progress In Electromagnetics Research C,* vol. 84, pp. 161-174, 2018.
[http://dx.doi.org/10.2528/PIERC18040605]

[15] A. Rahim, P.K. Mallik, and V.A.S. Ponnapalli, "Fractal Antenna Design for Overtaking on Highways in 5G Vehicular Communication Ad-hoc Networks Environment", International Journal of Engineering and Advanced Technology (IJEAT). ISSN: 2249–8958, Volume-9 Issue-1S6, December 2019, pp 157-160,
[http://dx.doi.org/10.35940/ijeat.A1031.1291S619]

[16] A. Rahim, "Design and Analysis of Multi Band Fractal Antenna for 5G Vehicular Communication", P-P 26487-26497, Vol. 83: March/April 2020, Test Engineering and Management, ISSN: 0193-4120,

[17] R. Kubacki, M. Czyżewski, and D. Laskowski, "Microstrip antennas based on fractal geometries for UWB application", *22nd International Microwave and Radar Conference (MIKON),* pp. 352-356, 2018.*Progress In Electromagnetics Research C,* pp. 352-356, 2018.

[18] G.P. Mishra, and M.S. Maharana, "Study of Sierpinski Fractal Antenna and Its Array with Different Patch Geometries for Short Wave Ka Band Wireless Applications", *Procedia Comput. Sci.,* vol. 115, pp. 123-134, 2017.
[http://dx.doi.org/10.1016/j.procs.2017.09.085]

[19] Al-saif H., Usman M., Chughtai M.T, and Nasir J., "Compact Ultra-Wide Band MIMO Antenna System for Lower 5G Bands", *Wirel. Commun. Mob. Comput,* vol. 6, pp. 1-2396873, 2018.

[20] S. Li, Y. Liu, L. Lin, X. Sun, S. Yang, and D. Sun, "Millimeter-Wave Channel Simulation and Statistical Channel Model in the Cross-Corridor Environment at 28 GHz for 5G Wireless System", *2018 International Conference on Microwave and Millimeter Wave Technology (ICMMT),* pp. 1-3, 2018.
[http://dx.doi.org/10.1109/ICMMT.2018.8563957]

[21] T. Mondal, S. Maity, R. Ghatak, and S.R.B. Chaudhuri, "Compact Circularly Polarized Wide-Beamwidth Fern-Fractal-Shaped Microstrip Antenna for Vehicular Communication", *IEEE Trans. Vehicular Technol.,* vol. 67, no. 6, pp. 5126-5134, 2018.
[http://dx.doi.org/10.1109/TVT.2018.2824841]

[22] V.A. Sankar Ponnapalli, and P.V.Y. Jayasree, "Thinning of Sierpinski fractal array antennas using bounded binary fractal-tapering techniques for space and advanced wireless applications", *ICT Express,* vol. 5, no. 1, pp. 8-11, 2019.
[http://dx.doi.org/10.1016/j.icte.2017.12.006]

[23]   A. Arif, M. Zubair, M. Ali, M.U. Khan, and M.Q. Mehmood, "A Compact, Low-Profile Fractal Antenna for Wearable On-Body WBAN Applications", *IEEE Antennas Wirel. Propag. Lett.,* vol. 18, no. 5, pp. 981-985, 2019.
[http://dx.doi.org/10.1109/LAWP.2019.2906829]

CHAPTER 3

Role of Artificial Intelligence in Market Development and Vehicular Communication

Arshi Naim[1,*]**, Fahad Alahmari**[1] **and Abdul Rahim**[2]

[1] *Department of Information Systems, College of Computer Science, King Khalid University, Abha 62529, Saudi Arabia*

[2] *Lovely Professional University, Punjab, India*

Abstract: In the present situation, applications of Artificial Intelligence are exercised by most of the service sectors for obtaining several benefits. This research article provides a general role of AI in financial services and its definition in practical scenarios. The research work is mostly qualitative and does not include any data analysis for its results. This chapter describes the major benefits derived by the practices of AI in general in the financial sector and vehicular communication. For the study, some examples of the role of AI in service organizations are analyzed and results are given from their findings. The authors have explained the benefits of AI from the perspectives of both customers and service providers. This research report provides brief information about AI in general and the advantages of AI for financial institutions in achieving its financial outcomes and helping in market development.

Keywords: Artificial Intelligence, Financial Institution Market Development, Financial Services, Financial Stability.

1. INTRODUCTION

Artificial intelligence (AI) has been accepted for various applications and benefits by industries for Financial Services (FS). Therefore it is imperative to examine the role and advantages of AI in a financial context such as how AI helps in achieving Financial Stabilities (FSt). The use of AI in FS is growing significantly [1]. The range of AI's implication varies; AI works for supply and demand factors in FS. The supply factors include advancements in technology, and for demand, it explains opportunities for profitability and level of competition. There are many applications of AI applied in present financial sectors that include [2, 3]:

---

[*] **Corresponding author Arshi Naim:** Department of Information Systems, College of Computer Science, King Khalid University, KSA. E-mail: arshi@kku.edu.sa

Praveen Kumar Malik, Pradeep Kumar, Sachin Kumar and Dushyant Kumar Singh (Eds.)

- AI is used by Financial Institutions (FI) and vendors to automate client interaction, measure credit quality concerning price, and in the context of market insurance contracts.
- FI can identify scarce resources and analyze the influence of the market value of the firm with the help of AI.
- AI facilitates FI and other firms to evaluate Return on Investment (ROI) and establish the correlation between funds available, allocation of resources, and hedging of the funds.

Adopting any technology is not always beneficial; both advantages and disadvantages are associated with its use. AI has also given rise to some issues about handling and privacy of data. Some problems have arisen due to AI applications such as the question of efficiency in the processing of information, how certain decisions may lead to achieving stability in the financial system, *etc.* These decisions may include decisions for credit, Financial Markets (FM), insurance contracts, and interactions with customers [4]. Another issue is about more dependencies on emerging technologies for achieving FSt. Last but not the least, AI considers many irrelevant and not related factors for achieving and measuring FSt.

This research article includes an exemplary scenario of the role of AI in PricewaterhouseCoopers (PWC), India.

2. LITERATURE REVIEW

AI and Machine Learning are fast-paced growing technologies and have been adopted by diverse fields. This research deals with the role of AI in FS.

Father of AI John McCarthy defined AI as the science of making intelligent machines, especially intelligent computer programs [5, 6].

AI began its journey in the 1950s with the Dartmouth Summer Research Project on AI at Dartmouth College, USA [4]. There are not many firms working without the use of AI in recent years because AI systems deliver more creative services that are results-oriented. PricewaterhouseCoopers (PWC) is a UK-based second-largest service serving professional organizations in the world. It is operating in many countries and India is one of them. This firm has partners across the globe and it is a lead firm in AI technologies, Data and Analytics [7 - 9]. PWC India has been applying AI in broader financial decisions to individual decisions. PWC India elucidates an application of AI techniques at clients as well as the firm's level. AI has been adopted more by service industries such as FIs and hospitals, *etc* in developing markets. Service industries have started applying AI systems for several benefits such as fraud detection, diagnosis of diseases, illustrating and identifying side effects, detecting prognosis, *etc* [10 - 12].

3. DISCUSSION

In the current scenario, sharing and analyzing of information has taken a digital platform and AI has a critical role in its working. For instance, AI specifically uses algorithms to automate the processing of financial data for market development. AI develops a liaison between FS markets, resolves issues about buying and selling, and enhances decision-making processes. AI is an example of emerging technologies and FIs are extensively applying it in checking the financial data sets for developing infrastructure, creating reliable policies for market growth and financial trading [13, 14].

A market development strategy [15, 16] is defined as a product or service-based long-term policy to create needs in the businesses. Market development Strategy seeks AI's applications in achieving this objective. AI helps not only locally but marketing at the global level [17]. The empirical application of AI can be witnessed in financial market development. Some examples are in the exportation of licenses for different financial marketing firms, merging and assembling joint ventures, or direct investments at diversified levels. But the process of applying is not very simple because the financial market is inclusive of several crucial dependents and independent variables in the economic system. The nation's economy is based on the financial market which concludes that surpluses and deficits have to be examined under many financial consequences. Tools and techniques of AI have tried to solve and support the process. Fig. (**1**) explains more about the working part of the procedure. AI has also assisted financial markets in meeting their objectives such as augmenting the capability of the financial market to act efficiently as an intermediary, negotiator, and creation of needs. Some more instances of AI in facilitating financial markets on the supply side are to provide and allocate a wide range of financial instruments, offering choices of issuers, credit risks, *etc*. and the similar scenario at the demand side also where AI helps in making investment decisions to meet trends and demand levels eventually reducing the risk factors [18]. There are highly liquid financial markets have the ability to accommodate large and varied issuance of financial instruments with minimum price effect which can be swiftly transformed and exchanged at a reasonable cost by the support of AI's application in automating the timeline and adding values also AI help Financial institutions and banking sectors to lower the transaction cost, enhance the efficiency of routine operations and can predict reasonable interests targets. FI uses AI's applications for three major benefits at the demand level such as identifying the opportunities for cost reduction, managing risk, and increase in productivity. Table **1** below shows the level at which AI aids in FI.

Table 1. AI at Demand Level.

Artificial Intelligence		
Financial Institutions		
Opportunities for Cost Reduction	Risk Management Gain	Productivity Improvement

Every firm or FI clearly describes the reasons for using AI. For market development (MD), these techniques are used to reduce costs, manage risk, and increase productivity. Also, AI is used for enhancing decisions for developing new products, identifying the best services for clients, and developing effective connections between systems and staff.

For example, the firm, PWC, India has listed in its study that it uses cases of AI for enhancing financial security systems. The study also explains the methods through which AI systems benefit in FS by reducing human errors, automating the process, and contributing to financial analytics in decision-making processes.

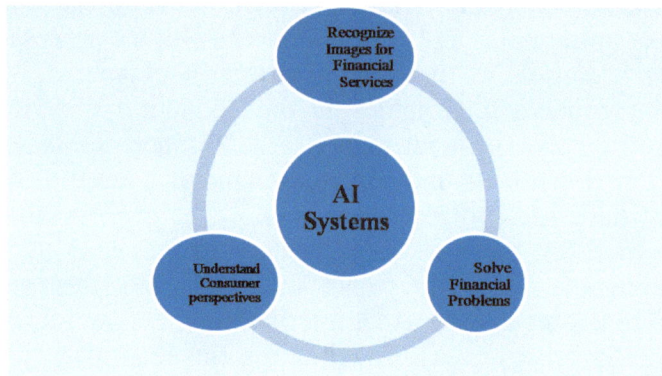

Fig. (1). Working Model of AI in PWC, India.

3.1. Micro Financial Point

FSt is dealt with at the micro-level and AI makes it possible in describing the factors that need to be considered for financial stability and recognizing the areas of improvement. Financial markets, consumers, and institutions are the players at a micro-level. Therefore it is vital to know how AI influences these parts. PWC, India annually prepares a report based on its financial results, consumers' survey, and stakeholder's satisfaction after using tools of AI in its working. PWC, India concludes the importance of AI at its micro-level in financial perspectives.

3.2. Possible Effects of AI in Financial Markets

AI has the prospects to affect the macro and micro levels in FM. It can process financial information accurately, reduce all information distortion, and make the financial system more efficient. Therefore financial markets depending on AI for processing of information achieve better results than without using it. PWC, India applies AI for analyzing information for market participants and for lowering market trading costs. Noticeably, AI affects service industries both at the micro and macro level.

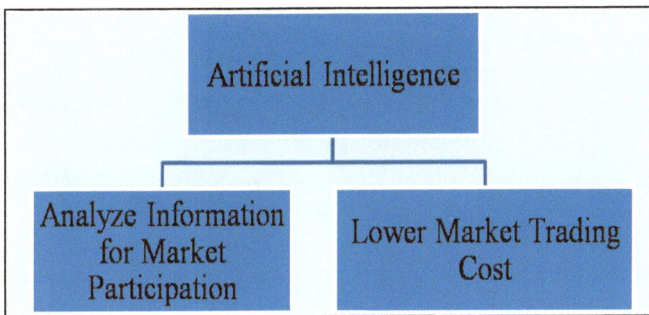

Fig. (2). AI in Financial Market.

AI techniques include the following in financial markets:

- AI uses several techniques in increasing financial market stabilities and development, like sentimental analysis to build up relationships with all key entities in FM, reduces all price irregularities, and develops effective pricing strategies as shown in Fig. (**2**). AI also assists in analyzing the information as a whole for the firm covering all major areas including financial decisions. For example PWC, India shows the role of AI in HRM, Financial decisions, Accounting decisions, user groups, *etc.*
- Most important reason for financial services to adopt AI is to lower market participants' trading costs that can comply with trading and investment strategies and develop markets for their consumers. AI also helps in discovering and reducing the prices over transaction costs.

The Financial Firm explains that AI can be used for credit scoring in FM. AI is used by financial sectors to explore opportunities and growth in businesses. There are some businesses also that are capitalized by AI in receiving benefits to produce quality data, accurate data, and cost efficiency [7 - 9]. These benefits encourage FIs to apply AI in their work to achieve financial stability and developing markets as shown in Fig. (**3**).

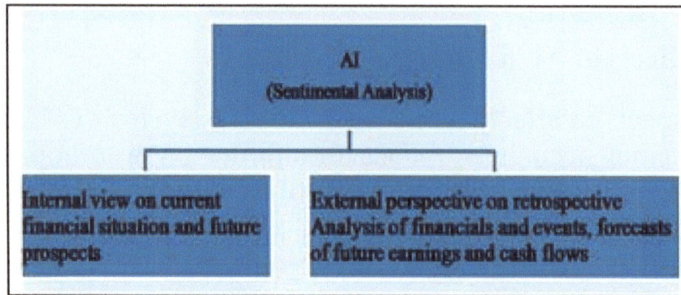

Fig. (3). AI in PWC, India [9].

3.3. Possible Effects of AI Learning on FI

It is assumed that if a firm is having large profitability consequently it will achieve financial stability. AI applications contribute to earning a good return on investment by controlling costs and monitoring other risk factors that prevent market growth.

The firm explains that AI applications automate the business processes that result in reducing routine costs. Also, the FI can use AI to know their customers, how to achieve customer satisfaction, resource allocation, and new product development. In the current scenario, AI techniques are more used by FI for detecting fraud, identifying suspicious transactions, and cyber attacks.

AI applications are used by FI for risk management too. This process is based on retrospective behavior of cash flows, various relations among prices of fixed assets, and revenue flow.

The firm describes that if financial decisions are mechanical by AI, they will offer accurate outcomes. Finance functions also indicate that AI is very impactful at all levels of operations.

3.4. Possible Effects of AI on Consumers and Investors

If AI reduces the costs and enhances the efficiency of FS, consumers can obtain several benefits.

AI aids customers in reducing borrowing costs increases the scope of financial services in their real life and can receive services and benefits based on their requirements. Another scope of AI is in big data analytics that can analyze the characteristics of customers that are available in the public domain. FI can take benefit of this analysis in understanding the consumer's needs and investors' requirements and accordingly can suggest financial products and services for them as shown in Fig. (**4**).

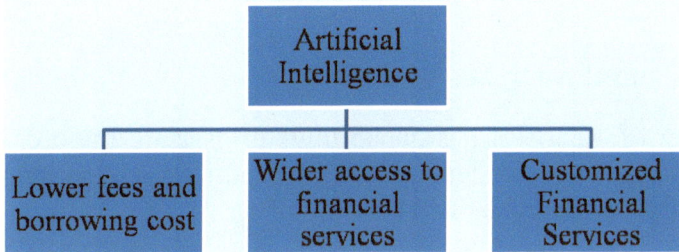

Fig. (4). Role of AI on Consumers and investors.

3.5. Possible Effects of AI on Vehicle Manufactures

As Vehicular communication is strengthening itself in the area of vehicle manufacturing, there is a direct impact of AI on the present technology. Vehicle communication is designed for the safety, security, and comfort of the users. As the communication is used for moving objects, there is always a possibility of disconnection in the communication as the vehicles are moving at rapid speed. Microstrip antennas are more efficient at fixed positions, but for moving objects, it becomes very difficult to detect the antennas and keep the communication stable. Search algorithms such as Depth-first search, Breadth-first search, greedy best-first search, A* search, and minimax algorithms can be effectively used for detecting the receiver. These algorithms use AI for increasing the probability of connecting the devices faster compared to the traditional algorithms. Manufacturers of vehicles are showing a lot of interest in AI-based algorithms such that the network can be stable and streamless communication is possible [19].

3.6. Possible effects of AI on Antenna Set up

AI has brought new dimensions in the field of electronic industry, the way it makes use of the resources to its best. One of the key components in any electronic gadget is the Antenna, as it the reason behind the communication in the wireless domain. The antennas are placed at a pre-defined position and the connectivity depends on the directivity of the antenna. The antenna is fixed at a position in the gadget due to which the directivity is constant. In moving object applications, it becomes tangent to make use of the antenna as the coverage depends on the position of the antenna. AI Knowledge-based algorithms are used to pre-determine the moving environment and adjust the antenna in a particular direction such that the signal can be maintained at constant connectivity. Industry 5.0 is advancing towards device-to-device communication in which the antenna position can be critically important. Knowledge-based search AI algorithms can be helpful in this domain as to set up the antenna with flexibility in nature and adjusted accordingly [20 - 21].

3.7. Macro-Financial Analysis

The application of AI to FS has capabilities to escalate the proficiency of the economy and contribute to understanding the need for the adoption of AI that can affect the financial systems on the whole.

Table 2. Impact of AI in the Financial System.

AI
Increasing the effectiveness of FS [IEFS}
Developing Alliance with Economies of Scope [DAES]
Stimulating investments through AI [SIAI]

Table **2** gives a clear picture that AI contributes to three imperative areas in financial systems namely, IEFS, DAES, and SIAI.

AI has contributed to providing benefits to clients; they can make effective financial decisions, allocate funds, and utilize them optimally as shown in Fig. (**5**). AI has made it possible to diminish the costs of payments for financial clients. AI has provided achievable standards for FIs by satisfying financial consumers with FS, by managing risk, liabilities, and loan portfolios as shown in Fig. (**6**). AI also elaborates the economies of scope for FI while realizing the relationship with other institutions in service sectors.

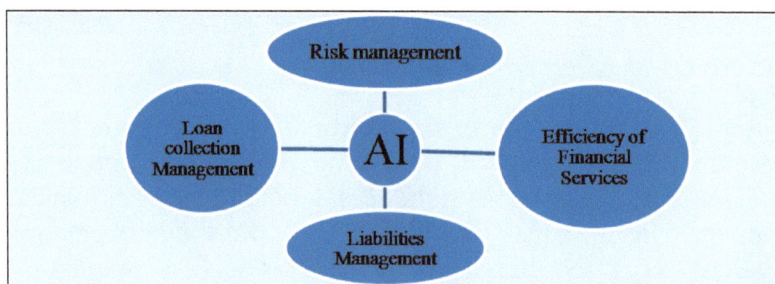

Fig. (5). Benefits of AI in Finance Systems.

This film explores how AI creates opportunities for financial and non-financial firms to grow together and get involved in research and development to stimulate economic growth. Macro financial aspects provide a range of benefits for the adoption of AI. It can be implemented from short-term to long-term economic growth.

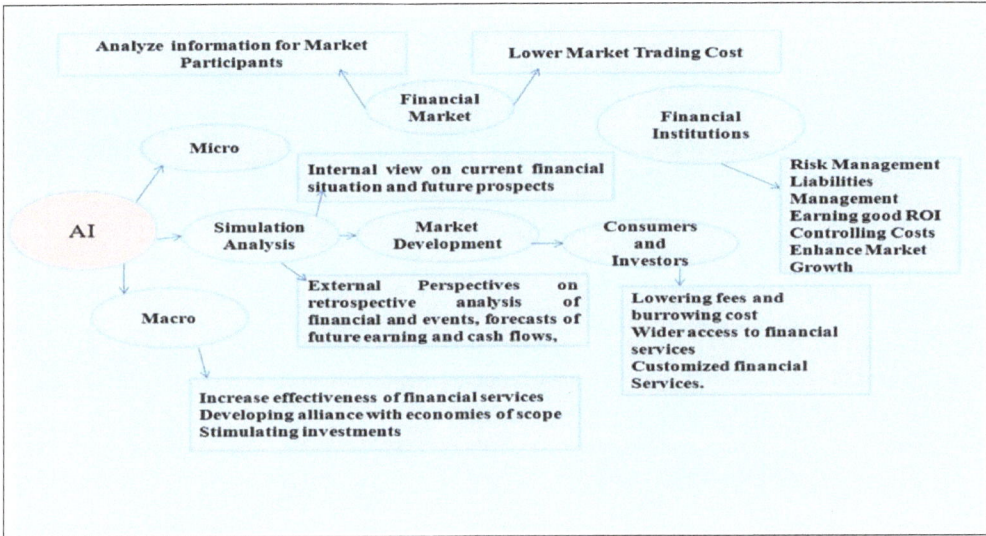

Fig. (6). Role of AI in FS.

This organization expresses the potentialities of AI in FS. AI focuses on financial systems from several perspectives, like how to stimulate economic growth so that FS using AI-powered solutions can provide relevant financial data and security systems. AI assists financial clients by informing new financial products, investment decisions, and solving financial problems. AI presents solutions to high costs and the lack of technical expertise. AI applications can integrate business processes, reduce costs, and bring higher productivity, growth, and efficiency.

4. RESULTS

This research article has concluded that AI provides key benefits to all service industries. There are many applications of AI used invariably by these service industries but this descriptive study explains the relevance and applications in Financial Systems and in developing the market in general.

Table **3** presents the comprehensive environment of AI application in FS. AI has been adopted to achieve benefits at various levels of micro and macro. This research article has illustrated the significance of AI and how it impetuses and exhibits the growth of the service industry in general and specifically for the financial sector. Since this paper deals with financial sectors we tried to identify the key benefits provided by AI for financial institutions and financial customers. Through the example of PWC, we presented the concern and benefits of AI in FS and for market development too.

The discussion part clearly explained how AI is playing a great role in PWC as an example and for any financial service providers in developing markets, achieving customers' satisfaction, enhancing security systems, controlling costs, managing risk factors, *etc*.

Table 3. AI Potentials in PWC, India [9].

Areas valued by AI	AI Potentials
AI stimulating Economic growth	Financial services using AI-powered solutions like a digital assistant, recommender systems, *etc* for helping financial clients and financial businesses for sharing data freely. Along with enhancing financial security systems.
AI as Customer care	Some systems of AI like chatbots, digital assistants, and robots can conduct and help financial clients, offer solutions to financial issues, give information on new financial products and services.
AI for market growth	AI applications can integrate business processes, reduce costs, and bring higher productivity, growth, and efficiency.

CONCLUSION

The use of AI will continue to be used more and more by service sectors and its various applications will make the services more autonomous, accurate, and cost-effective. Customer satisfaction and trust will be achieved by the service providers by using AI in their working and they will be able to provide more customized features too. Also, AI will play a more vital role in enhancing security systems, especially in the financial sectors.

CONSENT FOR PUBLICATION

Not Applicable.

CONFLICT OF INTEREST

The author confirms that this chapter contents have no conflict of interest.

ACKNOWLEDGEMENT

Declared none.

REFERENCES

[1]   Advantages and Disadvantages, issues of allocation of AI in FS "European Joint Committee Discussion Paper on the Use of AI by FI,".Jonathan Stuart Ward and Adam Barker (2016).,

[2]   "Definition of AI, Arthur Samuel (1959), for AI", *IBM Journal: 211-229; Tom Mitchell (1997), AI and Machine Learning,* vol. 349, Michael Jordan and Tom Mitchell: New York: McGraw Hill, no. 6245, pp. 255-26, 2015.

[3]   AI in algorithm, Thomas Cover and Peter Hart (1967), "Nearest Neighbor Pattern Classification", *IEEE transactions on information theory. hinking Machines: The Quest for Artificial Intelligence--and Where It's Taking Us Next,* vol. 13.1, Penguin Books: London, pp. 21-27, 2016. See Luke Dormehl

[4]   Dependencies in Emerging technologies, Tim Dettmers (2015), History of AI," Parallel Forall, December.2017, https://devblogs.nvidia.com/parallelforall/deep-learning-nutshell-history-training/)

[5]   AI in Assessing Risks: a Regulatory Perspective," OpRisk North America, June 21. Bauguess (2017). Gerard Hoberg and Craig M. Lewis (2015),

[6]   AI in FS and for FSt Brynjolfsson, E. & McAfee, A. (2017). The business of artificial intelligence. Harvard Business Review., https://hbr.org/coverstory/)

[7]   *Importance of AI in services PwC,* India, 2017. http://www.pwc.com/gx/en/issues/analytics/assets/pwc-ai-analysis-sizing-017/07/the-business-of-artificial-intelligence)

[8]   Policies of AI (2017), AI technology implecations, PWC, India Research Papers., http://www.nedo.go.jp/content/100865202.pdf)

[9]   AI used by Financial sectors, " explore opportinies, provide correct and reliable data" ' achieve cost effectiveness by AI" Application of AI Bauguess (2017),

[10] *AI in business, "effectiveness" "market development",* Rao, A, 2017. https://www.strategy-business.com/article/A-Strategists-Guide-to-ArtificialIntelligence)

[11] R. Kubacki, M. Czyżewski and D. Laskowski, "Microstrip antennas based on fractal geometries for UWB application", *22nd International Microwave and Radar Conference (MIKON),* 2018pp. 352-356. [http://dx.doi.org/10.23919/MIKON.2018.8405223]

[12] Market Development, achieving ROI, appropriate application of AI, results from the Survey PwC, India (2017), http://www.pwc.in/assets/pdfs/publications/ceo-survey/20th-ceo survey-being-fit--or-growth.pdfoklet.pdf)

[13] *New era of AI and machine learning 'Knight W,* 2017. https://www.technologyreview.com/s/609038/chinas-ai)

[14] *An decade of electronics, moving with AI in services PwC,* India, 2017. https://www.pwc.com/us/en/advisory-services/digital-iq/assets/pwc-digital-iq-report.pdf-awakening/

[15] AI for present scenario in FS and future of AI in service industries, Sara El-Hanfy,(Posted on:13 June 2018 – Categories): I, Industrial strategy challenge fund AI, Next Generation Services , https://innovateuk.blog.gov.uk/)

[16] V. Kumar, B. Rajan, R. Venkatesan, and J. Lecinski, "Understanding the Role of Artificial Intelligence in Personalized Engagement Marketing. Sage publications", *Calif. Manage. Rev.,* vol. 61, no. 4, pp. 135-155, 2019. [http://dx.doi.org/10.1177/0008125619859317]

[17] E.G. Popkova, and V.N. Parakhina, Managing the Global Financial System on the Basis of Artificial Intelligence: Possibilities and Limitations.*The Future of the Global Financial System: Downfall or Harmony. ISC 2018. Lecture Notes in Networks and Systems,* E. Popkova, Ed., vol. Vol. 57. Springer: Cham, 2019. [http://dx.doi.org/10.1007/978-3-030-00102-5_100]

[18] M. Ahmadi, S. Jafarzadeh-Ghoushchi, and R. Taghizadeh, *Presentation of a new hybrid approach for*

forecasting economic growth using artificial intelligence approaches., 2019. [http://dx.doi.org/10.1007/s00521-019-04417-0]

[19]   Al-saif H., Usman M., Chughtai M.T., and Nasir J., "Compact Ultra-Wide Band MIMO Antenna System for Lower 5G Bands", *Wirel. Commun. Mob. Comput,* vol. 6, pp. 1-2396873, 2018.

[20]   N. Shaik, and P.K. Malik, "A Retrospection of Channel Estimation Techniques for 5G Wireless Communications: Opportunities and Challenges", *International Journal of Advanced Science and Technology,* vol. 29, no. 05, pp. 8469-8479, 2020.

[21]   S.V. Akram, P.K. Malik, R. Singh, G. Anita, and S. Tanwar, "Adoption of blockchain technology in various realms: Opportunities and challenges", *Security and Privacy,* p. e109, 2020. [http://dx.doi.org/10.1002/spy2.109]

CHAPTER 4

Performance Analysis of Microstrip Patch Antenna for Various Applications

S. Kannadhasan[1,*] and **R. Nagarajan**[2]

[1] *Department of Electronics and Communication Engineering, Cheran College of Engineering, Anna University, Tamil Nadu, India*

[2] *Department of Electrical and Electronics Engineering, Gnanamani College of Technology, Namakkal, Tamil Nadu, India*

Abstract: An antenna is a leading transitional that is used to transmit or receive radio waves. In this ever-changing area of wireless networking, dual or multiband antennas have played a critical role in wireless service standards. The microstrip patch antenna has a number of advantages, including a cheap cost, a compact design, a simple construction, and optimum circuitry compatibility. The microstrip patch antenna is a popular and significant topic in antenna theory because of its numerous benefits over conventional antennas, including cheap cost, light weight, ease of feeding, and attractive radiation properties. It has been a source of frustration for scholars. an area of research in which the goal is to reduce antenna size while maintaining a high level of performance preserving a high gain and bandwidth For mobile phones and portable PCs, Microwave and Wireless Local Area Network (WLAN) Connectivity around the World WiMAX interoperability has been utilized. Both are affordable, efficient, and scalable as well as high capacity data connections. This article discusses a variety of applications for a variety of platforms. Radiating patches, feeding methods, and substrates come in a variety of shapes and sizes.

Keywords : 5G , Energy harvester , Multiband antenna , Neural Network , Textile Antenna and Smart antenna .

1. INTRODUCTION

The development of wireless communication technology such as computer, cellular technology, and individual area network for Remote Regeneration and Surrounding Information Observation needs an antenna that is suitable for dual or multi-band wireless communication devices. Broadband Antenna demands the

* **Corresponding author S. Kannadhasan:** Department of Electronics and Communication Engineering, Cheran College of Engineering, Anna University, Tamil Nadu, India;
E-mails: kannadhasan.ece@gmail.com and krnaga71@yahoo.com

Praveen Kumar Malik, Pradeep Kumar, Sachin Kumar and Dushyant Kumar Singh (Eds.)
All rights reserved-© 2021 Bentham Science Publishers

design of antennas in wireless communication areas. This needs to operate successfully over a wide range of frequencies. At the same time, multiband antennas are required for mobile communication technology that operate in different frequency ranges [1 - 5].

With the development of wireless communication systems, compact wireless devices have been needed which allow more space for the integration of other electronic components. Engineering innovation poses problems for integrating multiple RF-band antennas with a wide range of frequencies. We [CP1] can improve the antenna design by advancing the optimization methodology as well as provide us with the opportunity to analyze the existing studies to categorize and synthesize them in a meaningful way.

Since they are supposed to operate on a modern multi-band MIMO antenna, the new frequency band is required to cover any communication network with an increased data rate. The fifth-generation (5G) networking infrastructure is designed to use millimeter-wave bands with a vast amount of available spectrum to address the global bandwidth deficit in today's wireless cellular networks. Many tests have shown the promise of greater magnitude bandwidth orders combined with greater gain from beamforming and spatial multiplexing of multi-element antenna arrays. Consequently, developing an ideal antenna for the creation of millimeter-wave beams may be an important step in the realization of 5G wireless cellular networks [6 - 9]. Though there are many different beamforming techniques, the most common beam forming technology is the active-phase array. In designing and evaluating MIMO multiband, reconfigurable, faulty ground structure and metamaterial antennas, the antenna is to provide an overview of emerging research and development developments and creative approaches for 5G wireless applications. The design concept and methodology for the various Multiband MIMO technology, reconfigurable and DGS antenna has been described as operating in a single wireless network in a short-range communication and multiband antenna that covers different frequencies.

Low profile, low and multiple antennas play a significant role in targeting.tly attracted increased interest. Due to the need for multiband antenna to cover the amount of wireless network applications in less space, the demand for multiband MIMO Antenna Technology is growing for the current 4G and potential 5G. It is not a simple task to build the antenna, but special modifications or combinations of shapes should always be introduced or careful optimization is needed to get the frequency range of multiband. The 5G network would rely heavily on MIMO systems, as it needs limited space and lower costs. Such methods can be used to obtain all of these multi-band MIMO antennas needed [10 - 12].

The second approach is to use the self-like geometry in blue-printing antennas which are multiband or resonant over several frequency bands. Small antennas are of critical importance due to the space constraint on the device and the oncoming implementation and multi-input multi-output (MIMO) system. Nevertheless, the classical small antenna does suffer from inadequate efficiency. Fractal geometry offers the solution by designing compact and multiband antenna in the most effective and sophisticated way possible. There are several fractal geometries available, such as Sierpenski Carpet, Sierpenski Gasket, Koch Fractal Circle, Hilbert Curve, and the Contour Set [13 - 15].

The three important techniques for the design of the antenna are discussed in different applications. The efficiency of the multiband MIMO system needs to be improved by the mutual coupling between antenna elements in near spaces. The insulation methods in MIMO are difficult to compare because of individual characteristics such as operating frequency, protected area, *etc*. Some limitations in the present work are found below:

- Stark radiation distortion pattern
- Resonant-frequency shift
- Changes in impedance at the input

To achieve smart reconfigurability, certain techniques such as graph model, neural network, need to refine the antenna parameters and the design procedure. To achieve reconfigurability by using smart material, the antenna scale can be miniaturized. Using various bandwidth enhancement methods, such as using DGS and patch slot, will improve the antenna bandwidth. With different DGS the designs can be further improved to achieve good gain, performance, radiation pattern, current distribution, and depending on the application.

Many of the antennas reported drawbacks in the literature are designed for single band or dual wireless for the next generation.

2. DGS WITH BEAM STEERING CAPABILITIES OF 5G APPLICATIONS

The exponential growth of data traffic and mobile access has had significant and profound effects on the everyday life and social activities of people. Given the extremely rapid growth in wireless consumer goods and the prevailing development of the Internet of Things, the number of mobile networking devices is expected to reach 100 billion by 2020. Mobile information traffic also doubles (at least) each year, and by 2019 demand for multimedia or other data-hungry smartphone apps will surpass that for wired traffic. Although most individuals will receive 10 megabits (Mb / s) of 4G-mobile backed streaming in the future.

Modern cellular phone antenna construction is a demanding task on all accounts. Potential global 5G requirements and 5G network deployment require far smaller, more capable antennas. The antenna size will depend on the operating frequency and the necessary bandwidth. The International Telecommunications Union (ITU) has released a list of proposed 24–86 GHz frequencies (24.25–27.5, 31.8–33.4, 37–40.5, and 40.5–42.5 GHz) to match the uniformity of global mm-wave frequencies.

5G wireless antennas face greater challenges in the use of mm-wave frequencies, as the key considerations are versatility in design, efficiency and reliability in high bandwidth, and reduced multipath fading. Defected ground structure (DGS) components have been commonly used to improve the efficiency ojf filters, coplanar waveguides, microwave amplifiers, and working antennas.

The benefits of using DGS components are smaller part size, greater bandwidth, harmonic suppression, and excessive cross-polarization of the higher-order. DGS leads to disturbances that disrupt the uniformity of the ground and the stability of surface currents. DGS symmetrical structures act as resonant gaps which are inserted directly on either side of a microstrip line to effectively link the feed line.

DGS also modifies the shield current distribution of a defect in the ground to allow controlled excitation and the propagation of electromagnetic energy through the substratum, thereby altering the transmission line's capacitive and inductive response.

This means that DGS increases the effective capacity of an antenna or filter to generate a multiband antenna which results in the production of multiple resonant frequencies. The optimal resonant frequency can be tuned by choosing appropriate geometry and positioning at the correct locations for the installed antenna. Several wideband and multiband antenna configurations have already been reported, incorporating DGS as a symmetry/asymmetry into the radiating patch or ground level, and a single / period structure as slots or apertures.

For instance, a wideband metamaterial absorber was proposed in that it was independent of polarization and angle of incidence. Also, solar energy absorbers can be built using a meta-material structure formed in plus. The material properties of the triangular split ring can also be characterized by the resonator-based structure. Similarly, mushroom-shaped electromagnetic bandgap structures have been used for pattern reconfigurable low profile antennas. Several written studies suggest using antennas with mm-wave in one or more 5G bands.

3. TRI-BAND FOR 5G APPLICATIONS

It has become increasingly popular to use personal wireless devices (smartphones, smartwatches, smart glasses, *etc.*) that various frequencies. Such apps have many features, such as navigation, driving assistance, video capture, and accept/reject incoming calls, from a paired cell phone. In fact, these devices developed for emerging technologies are intended to work with the Internet of Things (IoT) definition at 5G (fifth generation) frequencies. The frequencies subject to this analysis have been selected so they fall within those intervals.

In addition, while compact antenna structures of feasible sizes are generally obtained by the use of coupling element structure in the literature, they have been obtained by the use of folded dipole and defective ground structure in this study. The SAR distribution of the tri-band antennas that radiate at Wi-Fi and 5G frequencies integrated in the eyewear system.

The SAR distributions in the human head are then analyzed for situations where the antenna is embedded in the frame and used alone because of the proposed antennas integrated into the 3D glass structure. The head phantoms of the homogeneous common anthropomorphic mannequin (SAM) and of the heterogeneous visible human (VH) were examined in both cases. The SAR values are assessed on the basis of the international standards.

4. MULTI-BAND ANTENNA FOR 5G APPLICATIONS

In the mobile device industry the demand for multifunction apps, such as smartphones, is growing. For eg, these gadgets have many latest mobile technologies: Wi-Fi, high-speed mobile internet (3G and 4G), GPS (Global Positioning System), GSM (Global Mobile System (Global Mobile Communication System), and Bluetooth. In a globalized world, it is a phenomenon that the Internet's velocity keeps rising over the years to conduct real-time video calls, among other functions, using a link given by a mobile operator, for example. In addition, these devices' displays are also growing in size for a better user experience with Hi-Fi video calls. Therefore, to reduce a smartphone's size, its internal components need to be shrinking. So, certain methods are used, such as designing integrated circuits, which greatly reduces the size of the circuit.

The need for a higher-speed mobile internet has stimulated innovation and the continual advancement of new technology such as 4G, which can achieve data speeds of 100 Mbps. Despite the fact that this technology has not been completely applied in many countries, there is a small rise in fifth-generation research and development, *i.e.* 5G mobile telecommunications, capable of operating at exceptionally high data transmission levels and utilizing electromagnetic spectrum frequency bands above 25 GHz.

Furthermore, the prospect of growth in the number of devices connected simultaneously after the introduction of the Internet of Things (IoT) worried mobile technology companies, which have found a solution efficiently to meet potential demand, emphasizing the value of the 5G rollout.

Basically, the microstrip antenna consists of three layers: a metallic layer above a substratum above another metallic layer, the ground floor. This third layer (ground plane), called configurations, may have different design formats. These are the parameters, which will, for example, determine the device's operating frequency.

Millimeter-wave is the frequency band between microwaves and IR waves, popularly referred to as extremely high frequencies (EHF) that are well suited to 5G networks. They require the transmission of frequencies from 30 GHz-300 GHz. Wavelength is within a range of 1 mm to 10 mm. The millimeter wavelets you take in more bandwidth. Wireless data are around 1 GB/s in the microwave frequency range and lower than those. The data velocity in the mm-wave frequency spectrum will reach up to 10 Gbit/s and more.

In the era of communication system growth planar microstrip antennas, 3G to 5G structures are very typical. They play a critical role in increasing the bandwidth and making extra narrow and modern touch suits. Microstrip lines or microstrip antennas are essentially a component for developing the planar geometry concept. Microstrip antennas are more effective and are used simply because they are easy to produce and versatile to vary their structure according to need and usage. Such patch antennas can radiate high-frequency waves. Such antennas can operate at frequencies of multiple and single bands.

5. ENERGY HARVESTER

As the demand for power increases, alternative energy sources became necessary. Electricity from natural sources such as solar power, thermal energy, wind power, and RF energy has been extracted for various purposes over the past few decades. Energy harvesting uses inexhaustible sources with little or no negative environmental effects and can provide ample energy for the lifetime of electronic devices. The research discussed here focuses on RF energy harvesting and the abundance of RF energy from nearby sources such as cell phones, wireless LANs

(Wi-Fi), FM / AM radio, and TV broadcasts The receiving antenna collects and rectifies signals.

Cell phones provide a significant source of transmitters to harvest RF energy from and potentially allow users to provide power-on-demand for a variety of near-sensing applications. Also relevant is the number of Wi-Fi routers and wireless endpoints such as laptops. A limited amount of energy (microwatts) can be collected from a typical Wi-Fi router with a 50–10 power level to receive RF energy from mobile base stations and broadcast radio towers in a practical manner for longer-range coverage.

RF energy harvesting is an attractive way to provide energy when energy sources are low and costly. Continuous propagation of electromagnetic waves from the base station, transmission tower, television, and devices such as mobile phones and radio results in excess of the dispersed electromagnetic waves in the surrounding environment. These waves can be harvested *via* an antenna system that converts the electromagnetic waves to useful electrical power. In this analysis, the microstrip patch antenna is preferred for the planned use of an RF energy harvester.

Microstrip patch antennas have been widely used in communications applications such as navigation, remote sensing, and telemetry because of their planar, low manufacturing cost, flexibility, integrated functionality, and enormous CAD software design and development capabilities. Incorporated comprehensive systematic research into microstrip patch antenna to improve the characteristics of radiation and increase the bandwidth. A microstrip patch antenna printed on high-resistivity or highly effective dielectric constant (nearly 10) substratum shows the degradation of surface wave excitation efficiency. Thus, lack of bandwidth, lower quality of radiation, and decreased patterns of radiation from the antenna.

The micromachined antenna is an idea of the characteristics of radiation that a micromachining technique achieves in a given thickness ratio to create an air cavity of the Si substrates. More specifically, the radiating patch is printed on a dielectric section composed of air and Si substrate. Micro-electromechanical system manufacturing technology (MEMS) has contributed to the development in micromachining. The benefits of MEMS micromachined construction are that it provides efficient radiation, wider bandwidth, and the full antenna system and other structures can be integrated and mounted on the same Si substrate.

The user density is rising day by day there is a major difference between the current user density now and a couple of years ago because each year a large number of new users get access to the internet, and there is a prominent growth in the wireless technology environment, as the user density is aggregate similarly higher data rates.

6. METAMATERIALS

Commercial UWB systems need antennas with a small bandwidth and low cost. It's a well-known fact that physical features like simple structure, small size, and low cost are very appealing. Limited bandwidth is therefore the main drawback of the microstrip patch antenna. Some techniques have been developed for improving the bandwidth. Such techniques increase substratum thickness mainly by using different shaped slots or radiating holes, stacking different radiating elements of laterally or vertically loading antenna, using magnetic dielectric substrates, and ground plane engineering as EBG metamaterials.

Model techniques were developed to produce antennas with specific properties that are capable of meeting the requirements in various applications. In some cases, the currently built antenna may satisfy those specifications, while the others may not. For example, certain antenna designs may meet impedance bandwidth specifications but the gains or sizes are not very good. Often the specifications are exceedingly difficult or practically impossible to meet. Antenna architecture techniques however continue to evolve to meet operational requirements.

A rectenne is a particular type of antenna which rectifies incoming electromagnetic waves into DC current. The development of rectennas for wireless power transmission and solar space transmission has achieved considerable success in implementing various functions and applications, such as RFID tagging systems, sensor batteries or condensers, WLANs, WiMax, and cognitive radio systems, as well as in medical applications. A typical rectenna consists of four main components: antenna, pre-correction filter, rectification filter for the circuit and pass filter for the DC. A microwave antenna is used to gather incoming RF power.

A conventional rectene consists of a dipole element or a dipole mesh that captures microwave energy and a Schottky diode for the rectification process. Several forms of rectenne elements have been proposed in the last few years. For example, the antenna can be of any type, dipole, Yagi-Uda, microstrip, monopole, loop, coplanar patch, spiral, or even parabolic. The rectenna can also take any type of rectification circuit, such as a full-wave shunt rectifier, a full-wave bridge rectifier, or other hybrid rectifier configuration.

The most common rectifying circuit in a serial configuration is a single diode, which also functions as a half-wave rectifier. We can also use a parallel half-wave rectifier, a tension doublers system to theoretically double the DC voltage output, or a dual-diode full-wave rectifier to improve conversion efficiency. Considering that a primary task for rectenna is to convert RF energy to DC power, the main design challenge is to achieve high conversion efficiency and there are essentially two approaches to achieving this aim. The first option is to obtain the full power and send it to the rectifying diode while the second option is to remove the rectifying harmonics.

Among the various types of antennas used in rectennas, microstrip patch antennas are becoming popular for use in wireless applications due to their low profile, low weight, low production cost, and ease and inexpensive to manufacture using modern printed circuit technology, compatible with a planer and nonplanar surfaces. The other reason for the widespread use of patch antenna is the versatility of patch antenna when selecting a particular patch shape and mode with respect to resonant frequency, polarization, pattern, and impedance. Therefore they are ideally suited for use as embedded antennas in mobile wireless devices and portable devices.

Numerous mathematical models were created for this antenna, and its applications were applied to many other fields. An important contributing factor for the advancement of microstrip antennas is the increase in electronic circuit miniaturization brought about by broad integration developments. Because conventional antennas are often a bulky and costly part of an electronic system, photolithographic-based microstrip antennas are seen as an engineering advancement.

In its simplest forms, a microstrip patch antenna consists of a radiating patch on one side of a dielectric substratum with a ground plane on the other. The patch is typically made of a conductive material such as copper or gold and can take demand for wireless communications, and lightweight, portable wireless communications devices continue to evolve, and designers are experimenting with ways to make these devices even smaller. One of the most important components of any wireless communications network are the antenna which should be lightweight but still offer high performance, generous gain, and broadband coverage By examining the effective medium capabilities of the SR-SRR for the first time, it is thought to be used as a metamaterial unit cell for the purpose of generating the high refractive index (HRI) medium around the antenna, so that antenna could be miniaturized or worked at a lower resonance frequency.

7. ARTIFICIAL NEURAL NETWORK

Optimizing design parameters now, for a few days, is the most important consideration to achieve good computational efficiency. Many methods of design optimization are proposed in the literature, including the genetic algorithm (GA), particle swarm optimization (PSO), biogeography-based optimization (BBO), neural network, and more. ANN algorithm for the optimization of the UWB antenna. Feedforward Is preferred to use the Radial Based Feature Network (RBF). ANNs are a dispensation process /(algorithms) that are then freely modeled on a reduced scale by the cerebral mammalian cortex neuronal structure.

A broad ANN model may have several hundred or a few thousand processing units, while a mammalian brain has billions of neurons, thus increasing the extent of its overall activity and increasing behavior. ANNs are the best choice for the optimization of microwave circuit statistical architecture. Neuro models are much more effective in computational terms than EM models. If the learning data from a "good" model trained the neuro models by either EM simulation or calculation, the neuro models can obtain efficient and accurate optimization, and the design can also be obtained in the training area.

Satellite contact and radar requests have been rapidly growing in recent years. Researchers have begun to explore the possibility of higher frequency (> 10 GHz) communication services *via*; Ku, K, Ka bands, due to concerns about the frequency spectrum below10 GHz with various commercial and security applications. Another problem facing researchers is the need for high-efficiency, broad bandwidth, fast data rates, and portable communication networks for computer security. With wideband / ultra-wideband (UWB) microstrip antenna structures all these criteria can be easily met.

By loading the varactor diodes, the antenna will adjust the lower and higher bands of resonant frequencies to cover ISM and WiMAX bands. There is a lot of interest in designing internal antennas which have the desired characteristics of robustness and compactness. A highly tunable antenna is required to satisfy the demands of large range applications. Most of the papers in the above literature discussed the single-band tuning mechanism with limited coverage of the bandwidth, larger antenna size, low data rate, and that performance.

8. SMART ANTENNA

Advanced middle-ware data transfer is just a part of the chain for efficient data transmission. Another significant thing is the consistency of the wireless service. This efficiency is highly affected by external factors (*e.g.*, atmosphere that blocks signals, other signals that cause interference), but it can be enhanced by designing antennas that take into account the context in which the restricted system is to be used. Supply-chain is one of the industries that depend heavily on efficient wireless communication. Nowadays, automated supply-chain data transmission is one of the main pillars of effective Industry 4.0.

Both array designs perfectly cover the GPS bands with a miniaturized implantable array antenna but both designs don't cover the LTE bands and face considerable size. A miniaturized GPS antenna but the suggested antenna is hard to create and does not cover the UMTS or LTE bands. A miniaturized low-profile antenna that covers the LTE band but does not cover the GPS band. The goal of this research work is to create a novel antenna integrating GPS technology (*e.g.*, GPS, GLONASS, GNSS, GALILEO) and 4G technology, taking into account the characteristics of use cases. This antenna can replace two antennas (GPS and 4G) resulting in cost savings and improves communication significantly.

A fixed-port isolation dual-band MIMO antenna setup, where the T-shaped patch placed the top of the patch along with DGS on the ground achieves high insulation. A dual-industry, science, and medical (ISM) band MIMO antenna with two sickle-shaped radiating patches, faulty ground structures (DGS), and a microstrip feed line where good insulation has been created. To receive high insulation, the bands acquired to use a dual-band MIMO antenna with two C-shaped monopoly antenna components and a decoupling system. A multi-band printed MIMO monopole antenna with coplanar waveguide (CPW) feed to improve the bandwidth capacity of its U-inverted slots and meander line slots.

A dual-band printed diversity antenna is deployed to combat multipath fading and increase insulation, it is printed on a PCB board, and a design technique has been suggested. Using the complementary split-ring resonator (CSRR), in which the CSRR is placed on the ground to achieve high insulation rates, a dual-band MIMO antenna is installed. A highly insulated dual-band MIMO antenna is proposed in which the high isolation is attributable to the two transmission lines mounted on the top of the substratum and with DGS. A dual-band slot MIMO antenna is proposed, whereby a high separation between two antenna components is achieved by the use of decoupling slot structures.

A triple-band notch MIMO antenna is suggested where it is used to achieve band notches with faulty ground compact electromagnetic bandgap, which is lighter than the traditional EBG configuration A dual-band MIMO antenna is proposed with a multi-port where there are four ports in total in the proposed antenna. This suggests a dual-band MIMO antenna consisting of both an inverted f decoupling section and a meandering resonating branch for higher isolation. A compact triple-band MIMO antenna with CRLH (composite right / left hand) unit cell used to achieve triple-band radiation as well as quasi-unidirectional.

Nowadays, the development of modern telecommunications technologies leads to massive output and deployment of limited-scale mixed-signal high-speed circuits. The high-density integration of circuits results in complicated electromagnetic interference (EMI). Early identification and minimization of such interference in the design process to save the expense of overall development in order to comply with the electromagnetic regulation. It has two types of Far-field and Near-field Probe. Using horn antenna or patch antenna, remote RF probe technique that includes identifying the source and power of the magnetic field.

As the sample size is large and the wavelength is low so it can not accurately measure the field from a small source, the measurement of the magnetic field is best suited to calculate the strength of the magnetic field of radio frequency(RF) from a small source. Loop samples are typically used to find the source and determine the magnetic field intensity in the area surrounding it. The principal concept is to use a loop wire antenna as a near-field probe to detect the field source from the circuit. The probe is then miniaturized into a basic low-cost loop made of a printed circuit board (PCB) without having to shield electromagnetically. The drawback is that this type of probe can't isolate unwanted couplings

On the other hand, the magnetic probes with electrical shielding offer better suppression of unwanted electrical field and better spatial resolution, while multilayer thick film is produced on silicon and glass substrates, thin film on a glass substratum, and co-fired multilayered ceramics. The thick film and thin-film sensors on the glass are brittle and can easily be damaged when measuring, thus offering a significant advantage in less side-electric field coupling. . The electric shielded magnetic probe can also be designed with low-cost glass epoxy laminated material or F-4 PCB, but is much larger than the thick film, thin film, and LTCC.

Nevertheless, they are both small band probes. Using the low microwave loss substratum, a near-field probe with a higher frequency of up to 20 GHz was recorded, but the low loss microwave PCB is much more costly than the FR-4 PCB. FR-4 material loss is higher than the microwave substratum and the FR-4 PCB probe results in lower magnetic field sensitivities.

9. TEXTILE APPLICATIONS

The rapid growth of wireless power transmission has increased the need for lightweight textile antennas that are high-gain and broadband-based. To create a lightweight antenna for wireless power transmission, we need to maintain as high antenna efficiency as possible for optimum power output to be achieved. Thanks to their cost-effectiveness, low profile, low weight, and quick implementation process, microstrip patch antennas are very advantageous. There are several ways to increase antenna bandwidth, including increased substratum thickness, antenna capacity, use of low dielectric substratum, use of various techniques to suit impedance, and feed.

The production of wearable computer device technology has been evolving rapidly to enhance the quality and productivity of human life by providing portable mobile systems. Significant prerequisites have been the accessibility, reliability, longevity, protection, and comfort of a wearable device. The device, known as intelligent clothing, knowledge transmission, and communication, has been used in medical area health monitoring operations, military surveillance systems, and disease prevention, and citizen medicine in the healthcare sector.

Textiles offer a good forum for the production of intelligent textile systems that combine antennas and electronics. Since the textile antenna has strong versatility and can be adjusted to everyday use. A communication device based on Smart Textile Antennas can be completely incorporated into the various clothes.

The wearable intelligent textile device in the field of application-oriented goods is an innovative fast-growing market. Enhanced communication and electronic technologies have allowed the development of lightweight, versatile, and intelligent smart antenna devices that can be installed on or implanted within the human body. Wearable smart textile antennas are classified as a low-profile hard dielectric substratum antenna and a small area of the flexible antenna with felt substratum or different textile antennas.

These textile-based devices exhibited strong electrical efficiency and mechanical versatility. But most of the recorded electronic textiles so far are thin metal wires, metallic tapes in clothing, or sputtered nanoparticles. The textile material used in jeans has a dielectric constant of 1.6 and a tangent loss of 0.0019 which gives the antenna a return loss of -19.62 dB to work at a 2.46 GHz frequency.

For the design of a patch smart textile antenna, a suitable variety of conductive material and non-conductive textile material is needed. Copper is selected and applied in the conductive materials for both patch and ground plane while the non-conductive jean textile is added to the antenna substratum, for antenna substratum jean is selected due to its good thickness and less permittivity of 1.6 which has good properties of textile antenna design.

An SMA connector excites the feed-line of the 50 ohm microstrip. The antenna's overall design characteristics contribute to other wearable features of the antennas. Some of the characteristics of the textile materials are given as Moisture absorption The moisture significantly changes the antenna's performance parameters when a fabric antenna absorbs water since water has a very high dielectric constant compared to the material. The fibers continually exchange water molecules with the air and change dynamic equilibrium with temperature and humidity in the surrounding air.

The various journal published in microstrip patch antenna is shown in Tables **1** and **2** and Fig. (**1**).

Table 1. Number of Paper Published in Peer Reviewed Journal.

Year	Number of Papers Published				
	IEEE	**IET**	**Hindawi**	**Wiley**	**ACM**
2010	48	48	59	30	30
2011	52	52	62	40	35
2012	56	56	65	50	40
2013	52	52	70	52	42
2014	58	48	80	38	45
2015	60	49	85	45	48
2016	62	52	86	60	49
2017	65	57	84	55	52
2018	68	60	86	40	57
2019	70	62	88	42	60
2020	70	48	90	48	62

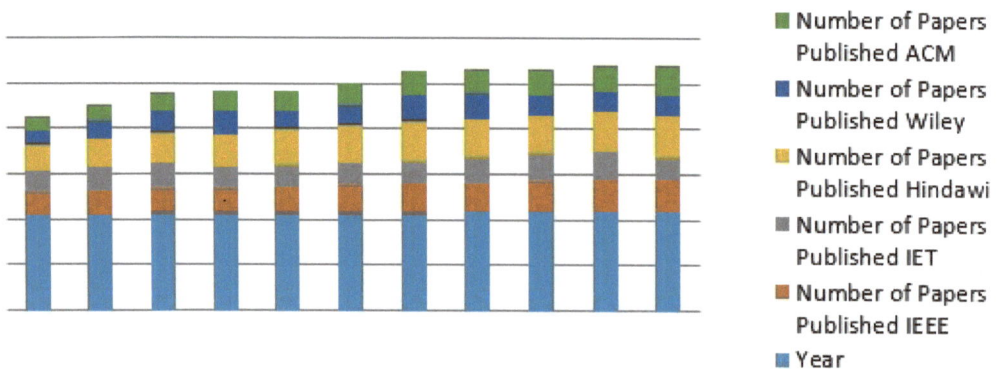

Fig. (1). Number of Paper Published in Peer Reviewed Journal.

Table 2. Number of Papers Published in Microstrip Patch Antenna.

Year	Number of Papers Published				
	5G Applications	Textile Applications	Multiband Antenna	Metamaterial Antenna	Smart Antenna
2010	10	12	20	30	16
2011	12	14	30	40	18
2012	20	16	35	50	20
2013	25	17	40	60	24
2014	28	12	46	52	26
2015	30	16	50	62	35
2016	15	18	48	55	40
2017	32	20	52	48	48
2018	18	18	38	50	50
2019	30	22	50	48	48
2020	20	25	58	45	50

CONCLUSION

This paper provides a theoretical survey on the microstrip patch antenna. It concluded that lower gain and poor power handling capability can be resolved by an array setup and slotted patch on the previous work in microstrip antennas after reviewing separate research articles. By utilizing several microstrip antenna topologies, microstrip antenna-based composite antenna formation, and sophisticated machining methods for microstrip antennas to strongly combine higher frequency antennas, these drawbacks can be reduced by concurrently juggling these three parameters. This paper provides an overview of the microstrip antenna and recent microstrip patch antenna creation for a broad variety of applications.

CONSENT FOR PUBLICATION

Not Applicable.

CONFLICT OF INTEREST

The author confirms that this chapter contents have no conflict of interest.

ACKNOWLEDGEMENT

Declared none.

REFERENCES

[1]   K.F. Tong, and T.P. Wong, *Circularly Polarized U-Slot Antenna IEEE Transactions on Antenna and Propagation,* vol. 55, no. 8, pp. 2382-2385, 2007.
[http://dx.doi.org/10.1109/TAP.2007.901930]

[2]   K.F. Lee, K.M. Luk, L. Tongy, Y.L. Yung, and T. Huynh, Experimental study of the rectangular patch with a U-shaped slot IEEE International Symposium Dig., 10-13, 1996,

[3]   J.A. Ansari, P. Singh, and N.P. Yadav, "Analysis of shorting pin loaded half disk patch antenna for wideband operation", *Progress in Electromagnetic Research C,* vol. 6, pp. 179-192, 2009.
[http://dx.doi.org/10.2528/PIERC09011203]

[4]   Rani, and R.K. Dawre, "Design and Analysis of Rectangular and U slotted Microstrip Antenna for satellite Communication", *International Journal of Computer Applications,* vol. 12, no. 7, 2010.

[5]   X.J. He, Y. Wang, J.M. Wang, T.L. Gui, and Q. Wu, "Dual-band terahertz metamaterial absorber with polarization insensitivity and wide angle", *Prog. Electromagnetics Res.,* vol. 115, pp. 381-397, 2011.
[http://dx.doi.org/10.2528/PIER11022307]

[6]   K. Kumari, N. Mishra, and R.K. Chaudhary, "An ultra thin compact polarization insensitive dual band absorber based on metamaterial for X-band applications", *Microw. Opt. Technol. Lett.,* vol. 59, no. 10, pp. 2664-2669, 2017.
[http://dx.doi.org/10.1002/mop.30797]

[7]   G.D. Wang, J.F. Chen, X.W. Hu, Z.Q. Chen, and M. Liu, "Polarizationinsensitive triple-band microwave metamaterial absorber based on rotated square rings", *Prog. Electromagnetics Res.,* vol. 145, pp. 175-183, 2014.
[http://dx.doi.org/10.2528/PIER14010401]

[8]   S. Ghosh, S. Member, and S. Bhattacharyya, "An ultra wideband ultrathin metamaterial absorber based on circular Split rings", *IEEE Antennas Wirel. Propag. Lett.,* vol. 14, pp. 1172-1175, 2015.
[http://dx.doi.org/10.1109/LAWP.2015.2396302]

[9]   H. Oraizi, and N. Valizade Shahmirzadi, "Frequency- and time-domain analysis of a novel UWB reconfigurable microstrip slot antenna with switchable notched bands", *IET Microw. Antennas Propag.,* vol. 11, no. 8, pp. 1127-1132, 2017.
[http://dx.doi.org/10.1049/iet-map.2016.0009]

[10]   N. Tuffy, L. Guan, A. Zhu, and T.J. Brazil, "A simplified broadband design methodology for linearized high-efficiency continuous class-F power amplifiers", *IEEE Trans. Microw. Theory Tech.,* vol. 60, no. 6, pp. 1952-1963, 2012.
[http://dx.doi.org/10.1109/TMTT.2012.2187534]

[11]   S. Kannadhasan, and R. Nagarajan, "Performance Design and Development of Environmentally Safe W-Shaped Structure Antenna for Wireless Applications", *Journal of Green Engineering (JGE),* vol. 10, no. 9, pp. 4558-4565, 2020.

[12]   M. John, and M.J. Ammann, "Wideband printed monopole design using a genetic algorithm", *IEEE Antennas Wirel. Propag. Lett.,* vol. 6, pp. 447-449, 2007.
[http://dx.doi.org/10.1109/LAWP.2007.891962]

[13]   S.K. Oh, H.S. Yoon, and S.O. Park, "A PIFA-type varactor-tunable slim antenna with a PIL patch feed for multiband applications", *IEEE Antennas Wirel. Propag. Lett.,* vol. 6, pp. 103-105, 2007.
[http://dx.doi.org/10.1109/LAWP.2007.893096]

[14]   S. Kannadhasan, and R. Nagarajan, " Performance Design and Development of Environmentally Safe W-Shaped Structure Antenna for Wireless Applications", *Journal of Green Engineering (JGE),* vol. 10, no. 9, pp. 4558-4565, 2020.

[15]   S. Shrivastava, and A. Bhargava, "A Comparative Study of Different Shaped Patch Antennas with and Without Slots", *Int. J. Eng. Dev. Res.,* vol. 2, no. 3, pp. 3306-3312, 2014.

CHAPTER 5

Importance and Uses of Microstrip Antenna in IoT

Wani V. Patil[1,*]

[1] *Department of Electronics Engineering, G H Raisoni College of Engineering, Nagpur, Maharashtra 440016, India*

Abstract: This chapter attempts to give consideration to how the IoT (Internet of Things) is attempting to revolutionize the world. IoT is a network wherein different devices interconnect and furnish to communicate with one another, helping to drive computerization to an advanced level. This helps all the connected devices to communicate with one another and decide by themselves without human intervention. Unconsciously, IoT applications are turning into a significant part of our life. The IoT resembles the system comprising different objects like devices, instruments, vehicles, unique structures, and different things implanted with electronic circuits, programming, different sensors, and system-network connectivity which empowers all of them to collect information. The IoT permits these different objects to sense and control distantly across existing network structure, and thus enables the chances for the physical world to incorporate into computer-based systems, enhancing the effectiveness and precision.

Keywords: Antenna Performance, Internet of things, Microstrip antenna, Wireless Communication.

1. BIRTH OF INTERNET OF THINGS

We can elaborate IoT as the "Internet of Things". In 1999, "Kevin Ashton, in an introduction to Proctor and Gamble, a co-founder of MIT's Auto-ID Lab proposed The Internet of Things. He spearheaded RFID (used in a standardized identification /bar-code identifier) for the applications of the supply-chai--management system. In 1982, changing the Coke machine introduced the network of the smart-devices concept at Carnegie Mellon University into the first internet-connected appliance [1] with the ability to report about its stock and to inspect whether recently stacked-beverages were cold. Kevin Ashton is a British innovator who invented the term "the Internet of Things" by introducing a framework where the Internet is connected with the physical world by pervasive

---

[*] **Corresponding author Wani V. Patil:** Department of Electronics Engineering, G H Raisoni College of Engineering, Nagpur, Maharashtra 440016, India; E-mail: wani.patil@raisoni.net

Praveen Kumar Malik, Pradeep Kumar, Sachin Kumar and Dushyant Kumar Singh (Eds.)

sensors. In this way, Kevin Ashton made the first statement which was written in 2009 in the RFID Journal. If we had computers that knew everything there was to know, using data they gathered without any help from us, they could track and count everything, and reduce waste, loss, and cost. They would know when things needed replacing, repairing, or recalling, and whether these were fresh or past their best. We need to empower computers with their own means of gathering information, so they can see, hear, and smell the world for themselves, in all its random glory. Kevin's statement highlighted the ideologies regarding the improvement of IoT. In IoT, the word 'Thing' points out towards any device embedded with different sensors having the capability to collect and transfer data over a network with no manual intercession. The embedded technology helps the devices to interact with internal states and the external environment and thus helps in the decision-making process. More or less, IoT is a platform that connects different devices to the internet and allows all to communicate with one another over the internet. IoT is a system of all interconnected devices that assemble and share the data. As a result, each of the assembled devices will learn from the experiences of the devices connected in the network like the human being used to do. IoT extends the human association, *i.e.* connect, contribute and collaborate to things. However, the system is complicated and we can elaborate it with the help of the model discussed in the next part.

An engineer or developer presents any of his designed applications with the record including the logic of the principles, expected errors and exceptions designed by him to the tester. Then issues encountered by the Tester are forwarded to the developer which includes multiple iterations and in this way a smart application is developed. Even the sensor at room temperature accumulates the information and sends it over the network that likewise helps to change the temperature of the other multiple devices sensors. For example, the sensor included in the refrigerator accumulates the information regarding the outside temperature and respectively changes the refrigerator's temperature. Likewise, the air-conditioners can change their temperature accordingly. This is the way devices can communicate, contribute, and collaborate as shown in Fig. (**1**).

1.1. Advantages of the IoT

As the IoT enables the physical world to be controlled distantly by creating the opportunities to connect and coordinate with the computer-based frameworks over the internet by utilizing sensors and the internet, it results in automation in every field and furthermore empowers the advanced applications. The evaluation of such systems includes enhancement in accuracy, efficiency, and economic benefit without human interaction. IoT includes new technologies such as smart-grids, smart homes, intelligent transportation, and smart cities as shown in Fig. (**2**). The significant advantages of IoT are:

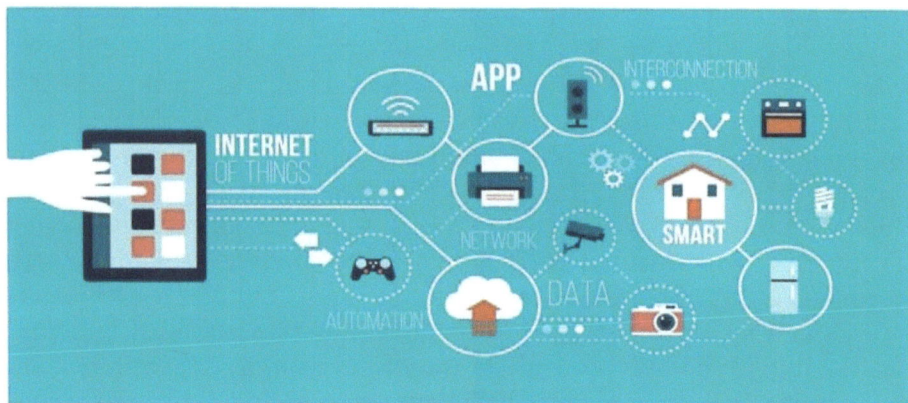

Fig. (1). What is IoT?

Fig. (2). Connecting multiple devices.

• Improvement in the Engagement of the Customer – IoT helps to improve the customer experience by using automated action. For example, the sensors incorporated in the smart vehicle can automatically distinguish the issues and can timely notify the driver and the manufacturer about it. Before the arrival of the driver at the service station, the faulty part can be made available by the manufacturer at the service station.

• Optimization of the Technical Approach– IoT promotes innovations and helps to improve them. The manufacture can improve their plan and try to make them more proficient by gathering information from various vehicle sensors and investigate them.

• Reduction in the waste – IoT uses real-time information for effectively processing the decision-making capacity for the effective resources management approach. For instance, if a manufacturer discovers faults in various engines, he may follow the assembling plant of those engines and can amend the issues with assembling belts.

These days, the world is surrounded by different IoT-empowered devices that are ceaselessly emanating information that is being communicated through different devices. Let us consider the necessary equipment used in the IoT hardware application. The applications of IoT can be found in every field of the communication system including the antenna design. An antenna is the most required part of communication devices like smartphones, smart cities, smartwatches, smart rings, smart TV, smart helmet, *etc.* The communication system must be efficient to transfer all types of data such as voice text and multimedia information. Microstrip antenna is broadly utilized in used specialized devices because of its low conformal and low profile properties [1]. Currently, industries are focusing on the Internet of things and their application. IoT devices more often required to communicate wirelessly as they can easily access the internet [2].

2. DESIGN OF MICROSTRIP ANTENNA FOR IOT APPLICATIONS

The IoT is an environment that promotes the information transfer capacity over the system network without using the human-to-human or human-to-computer interaction. Many research works are carried out in the microstrip antenna design for IoT applications which may incorporate communication-based applications. Many wireless communication applications use microstrip patch-antenna. For example, a circularly-polarised radiation pattern obtained either from a square or circular patch microstrip antenna is used in satellite communication systems nowadays. Microstrip antennae are also found applications in global positioning satellites (GPS). They are minimal in size and very costly because of their positioning. Some of the ongoing IoT applications utilizing the microstrip antenna are discussed following examples.

2.1. Multiband Microstrip Patch Antenna for IoT Applications

IoT is utilized for a short-range communication system that needs narrow bandwidth in most of the IoT applications [3]. Normally BW changes from

10MHz to 100MHz [4]. The specifications of the general IoT system include short-range, low power, omnidirectional, and seamless connectivity for a short time duration. Due to the increasing growth in IoT demand, new frequency-bands spectrum range from sub GHz to several GHz is worldwide released by the spectrum regulatory authorities. For this purpose, the antenna structure implemented in the IoT applications should fit in the defined range of frequency spectrum bands ie (830 to 840MHz), (850 to 890MHz), and (1190 to 1200MHz) range. This is accomplished by the tight narrow band criteria.

Planar Microstrip patch Antenna is designed that is dependent on the entrenched hypothesis of the design of Antennas for high frequencies (over a few hundred MHz) by simple planar structures [1]. Such designed microstrip patch antenna approaches to the PCB-like structures [5]. The PCB board consists of the dielectric material of glass epoxy (FR4) having the thickness h=0.1589cm sandwiched between a perfect dielectric conductor (PEC) like thin copper or other material on the two sides of the dielectric material. Fig. (3) elaborates on the physical shape, measurements, and feeding strategy for excitation [6].

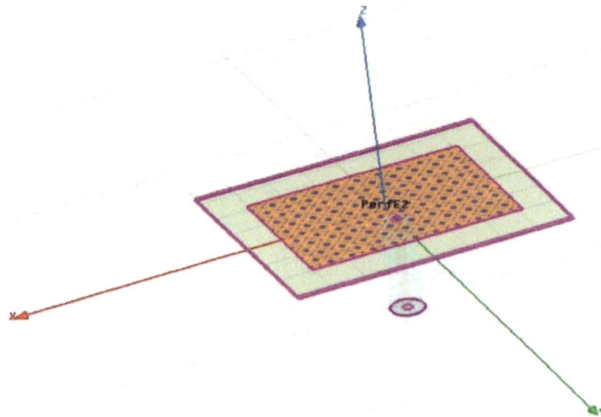

Fig. (3). Planar microstrip patch antenna [7].

By maintaining appropriate dimensions and feeding position, the exceptional element design can be achieved as shown in Fig. (**4**). This design is adequate to transmit or receive signals in specific frequency bands. Eq 1, Eq 2, and Eq 3 are required for the design of such a simple planar microstrip patch antenna, which is demonstrated as follows:

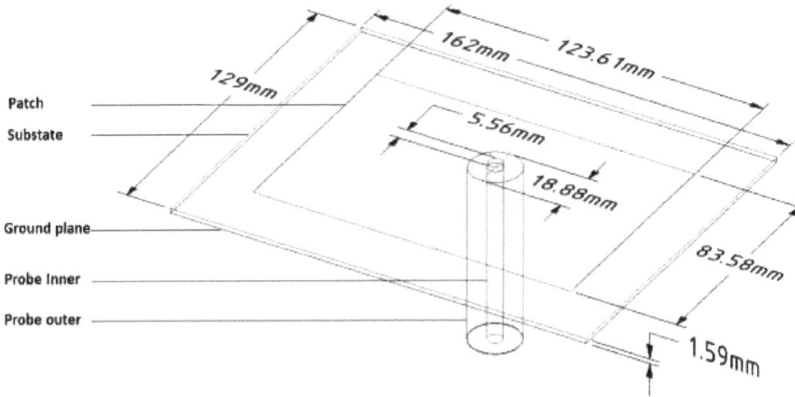

Fig. (4). Microstrip planar patch antenna.dimensions [7].

Table **1** elaborates antenna dimensions required for achieving the design specifications [1, 6] of the antenna.

Table 1. Key Antenna dimensions [7].

Name	Value	Unit	Evaluated Value
sub H	0.1589	cm	0.1589 cm
subX	16.2	cm	16.2 cm
subY	12.9	cm	12.9 cm
coax_outer_rad	0.944	cm	0.944 cm
feedY	1.43374	cm	1.4337 cm
gnd_x	subX	cm	12.9 cm
feedX	0	cm	0 cm
gnd_y	subY	cm	12.9 cm
patchX	12.36096	cm	12.36096 cm
patchY	8.35822	cm	8.35822 cm
coax_inner_rad	0.278	cm	0.278 cm
Feed length	6.105	cm	6.105 cm

Length of the patch is

$$L = L_{eff} - 2\Delta L = c/(2f_0 - \sqrt{\varepsilon_{eff}}) \qquad (1)$$

Where $\Delta L = h / \sqrt{\varepsilon_{eff}}$

$$W = \frac{c}{\frac{2\,f0\sqrt{\varepsilon_r+1}}{2}}$$

(2)

$$\varepsilon_{eff} = \frac{\varepsilon_r+1}{2} + \frac{\frac{\varepsilon_r-1}{2}}{\sqrt{1+\frac{10h}{W}}}$$

(3)

ΔL - fringing filed length along radiating edge of the patch.

h - substrate thickness

ε_{eff} - dielectric effective permittivity.

ε_r - substrate relative permittivity

W- patch.width

L- patch length.

c- free space velocity.

ε_o -free space permittivity *i.e* air.

f_o - antenna resonance frequency.

During the antenna design approach, the antenna specification concentrated on maintaining n specified narrow bandwidth, 10MHz at f_{r1} (m1), 40MHz at f_{r2} (m2), and 10MHz at f_{r3} (m3). Fig. (**5**) elaborates the frequency to the return loseS_{11}, required for obtaining the design specification.

For the achievement of the equivalent of the designed antenna, the IoT applications use omnidirectional radio wires which is satisfied by having practically uniform far-field radiations aside from at $\theta = 0°$, and $\theta = 180°$. The same is portrayed in Fig. (**6**).

The proposed work aims at designing a multi-band IoT antenna having a notch bandwidth of 10 MHz, 40 MHz, and 10 MHz resembling the approximate resonate frequencies 825 MHz, 870 MHz, and 1195 MHz by simulating in the HFSS programming environment. At the resonant frequencies, the return loss is optimized to -20 dB which is far better than the minimum requirement of -10dB. Subsequently, the specifications of the application are fulfilled by the performance of the antenna [7].

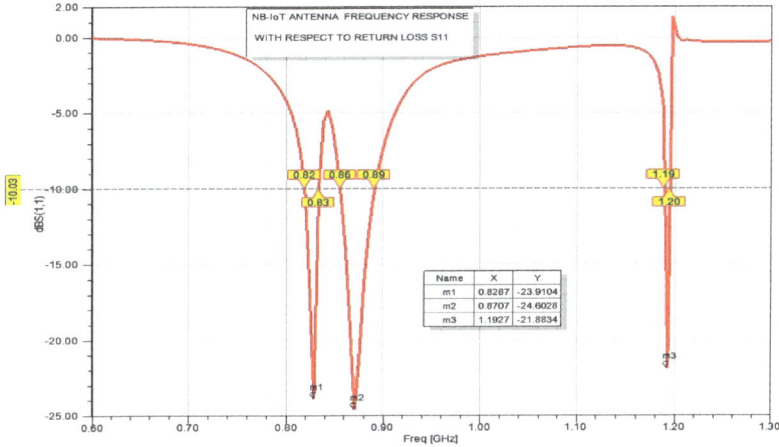

Fig. (5). Frequency to return loss S_{11} response in dB [7].

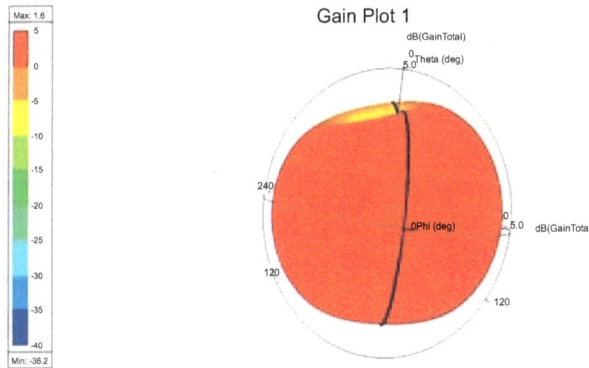

Fig. (6). Omnidirectional radiation pattern with gain variation [7].

2.2. Design of A Handy Triple band Micro- Strip Patch Antenna for Satellite-Based Iot Applications

A microstrip patch antenna is useful for high performance in extreme applications. Low profile, conformable, simple, and inexpensive, mechanically robust, and very versatile are the salient features of the microstrip patch antenna.

For high-performance applications, a microstrip patch antenna basically used a metallic strip or patch mounted on a dielectric layer over a grounded plane which is simple, conformable, inexpensive to manufacture, having a low profile along with mechanically robust and very versatile properties. The important properties of the microstrip antenna include low-efficiency low power and high-quality factor. Currently, most of communication devices like mobile phones, tabs, Wi-Fi modems [8] are using the microstrip patch antennas due to their low profile and conformability the microstrip patch antenna's performance relies upon the type of substrate utilized alongside the thickness of the substrate. Depending upon the permittivity of substrates, the amount of radiation fluctuates. Higher the permittivity, the higher the energy stored and the vice-versa. The performance of the antenna is evaluated by utilizing the parameters like return loss and voltage standing wave proportion.

The microstrip patch antennas are effectively found applications in handheld and minimized remote devices [9] due to their potential benefits of small size and lightweight. For achieving antenna devices with more conservative and high performance, the size of the antenna is diminishing exponentially to enhance the miniaturizing technology. Nowadays, the popular handheld wireless devices support the multiple band frequencies, as a result, the researchers are concentrating their work on downscaling the antenna in the absence of the impact of the reception apparatus execution. An enormous portion of them relies upon altering the fixed shapes, for instance, introducing slots, and blemished ground structures. Defected ground structure (DGS) [10] is a popularly utilized technology for the reception apparatus.

The slots etched relating to the ground plane are referred to as defected ground structures (DGS). In DGS, the defected structures might be single or multiple in number supporting the improvement of gain, operating bandwidth just as suppressing higher mode harmonics and shared coupling. At first. DGS was accounted for filters underneath the microstrip line. Later, the defective ground structures are utilized for acquiring stop band attributes/characteristics too. Subsequently, the DGSs continuously turned into powerful inclusions in altering the attributes of the microstrip antenna, for example, radiation pattern, operating bandwidth, gain, and the return-loss towards the requirement of the desired applications.

The proposed antenna is designed in HFSS microwave recreation programming utilizing Rogers RT/duroid 6010/6010LM as appeared in Fig. (**6**). At first, the antenna has been planned to utilize FR4 substrate for which, the acquired outcomes as far as return misfortune were extremely poor. As a remedy, the substrate has been replaced with Rogers RT/duroid 6010/6010LM and then simulated. Both the outcomes were compared and the results observed to be so strong that the observer is solid to such an extent that, the return loss parameter in the event of RT/duroid ($\varepsilon r = 10.2$) being the substrate has been over the standard of - 10dB at three distinct frequencies as appeared in Fig. (**7**). Though, the previous case failed to demonstrate that it is advantageous. The dimensions of the proposed antenna which is structured utilizing Rogers RT/duroid 6010/6010LM are indicated as appeared in Table **1**.The outcomes relating to the different receiving wire boundaries including the parameters of the antenna-like return loss, radiation pattern, and gain that have recorded.

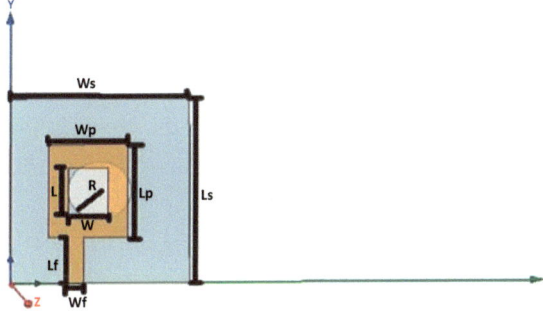

Fig. (7). Proposed design of the Antenna [9].

The proposed antenna works at three particular frequencies 3.05 GHz, 7.275 GHz, and 8.55 GHz in the ultra-wide band [11] region, where the possibility for interference between any two of the three working bands has been barely least because of the high dielectric substrate used. The higher permittivity of the RT/duroid substrate (εr = 10.2) comprised the 1comprehensive isolation between each of the operating bands, where the amount of return loss is significantly high. This specific quality of the proposed antenna has made it a distinct device that can be dedicatedly utilized for real-time screen sharing of satellite information. The patch and the slot measurements depend on the numerical expressions of width, length, and radius which are relying upon different significant parameters, for example, the relative permittivity of the EM wave, operating frequency, height, and width of the substrate. The design specifications of the proposed antenna have been extracted from the accompanying numerical expressions and shown in Table 2.

Table 2. Design Specifications of the Proposed Antenna [9].

Parameter	Substrate	Patch	Ground Plane	Feed Line	Slot	DGS (CIRCLE)
Length (L)	32mm (Ls)	16mm (Lp)	32mm (Lg)	8mm (L)	8mm (L)	-
Breadth (B)	28mm (Ws)	12.45mm (Wp)	28mm (Wg)	2.46mm (Wf)	6.225mm (W)	-
Heigth (H)	0.8 m (h)	-	-	-	-	-
Radius (R)	-	-	-	-	-	5mm (R)

$$\text{Width} = \frac{c}{2f_o\sqrt{\frac{\varepsilon_r+1}{2}}} \; ; \; \varepsilon_{\text{eff}} = \frac{\varepsilon_r+1}{2} + \frac{\varepsilon_r-1}{2}\left[\frac{1}{\sqrt{1+12(\frac{h}{W})}}\right] \qquad (5)$$

$$\text{Length} = \frac{c}{2f_o\sqrt{\varepsilon_{\text{eff}}}} - 0.824h \left(\frac{(\varepsilon_{\text{eff}}+0.3)(\frac{W}{h}+0.264)}{(\varepsilon_{\text{eff}}-0.258)\,(\frac{W}{h}+0.8)} \right) \tag{6}$$

$$a = \frac{F}{\{1+\frac{2h}{\pi\varepsilon F}\left[\ln\left[\frac{\pi F}{2h}+1.7726\right]\right]\}^{1/2}} \qquad \text{where } F = \frac{8.791 \times 10^9}{f_r\sqrt{\varepsilon}} \tag{7}$$

Parameters involved in designing of antenna

w - Width of antenna

h - Height of antenna

f_o - operating frequency

ε_{eff} - effective dielectric constant.

a - radius of the circular slot.

The comparative results of the return-loss parameter for designed antenna planned to utilize FR-4 as well as RT/duroid is as appeared in Fig. (**8**).

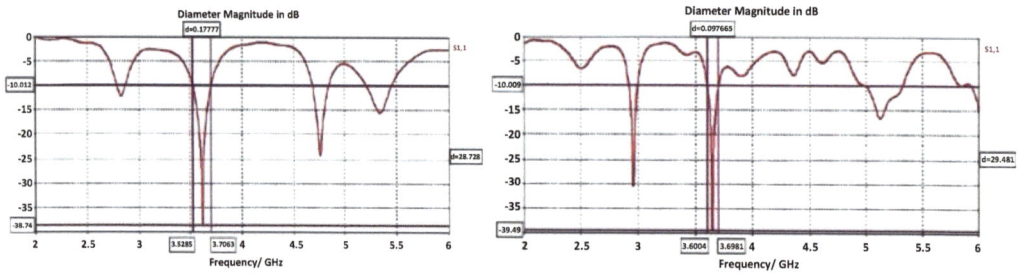

Fig. (8). Return loss in dB for (A) design using FR4 epoxy; (B) design using RT/duroid [9].

The outcomes/results clearly reveal that, the antenna designed utilizing RT/duroid is more efficient when compared with the designed antenna utilizing FR4 epoxy. For the designed rectangular-patch along with a rectangular-slot and the ground structure punctured with a circle, the impedance matching has been poor, and consequently with FR-4 ($\varepsilon r = 4.4$) as substrate, there is hardly a frequency at which the antenna will, in general, tends to work [13]. Then again the antenna with RT/duroid having exceptionally high permittivity ($\varepsilon r = 10.2$) in combination with impedance mismatch, acquired the tendency of radiating fields in spite of sustaining them through. With this change of impedance at separated frequencies inside the ultra-wide (3.1-10.6 GHz) region, the proposed antenna works at the three (3.05 GHz, 7.275 GHz, and 8.55 GHz) particular frequencies, when the height of the substrate (RT/duroid) is chosen as 0.8mm.

The resultant radiation pattern for the designed antenna is obtained at 3.05 GHz, 7.275 GHz, 8.55 GHz are as shown in Fig. (**9**).

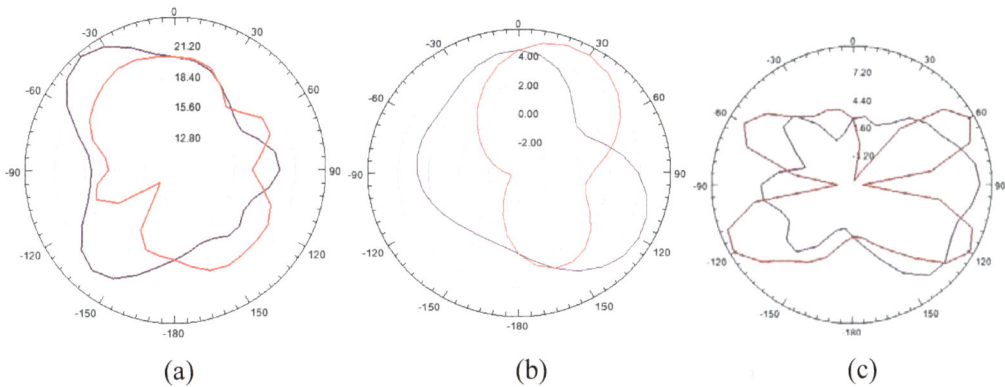

(a)   (b)   (c)

Fig. (9). Designed antenna Radiation Pattern at (A) 3.05 GHz (B) 7.275 GHz (C) 8.55 GHz respectively [9].

2.3. 3D Miniature Antenna Design for RFID Applications in IoT Environment

A framework that is used to transmit the identity of any type of objects, products, or any individual wireless utilized radio waves can be obtained by utilizing the RFID (Radio-frequency identification) [12]. For the last 10 years, a lot of innovative technology is applied in radio communication which is increasing rapidly and their performance is improved by reducing the size as well as the cost of the antenna and making them reliable. An essential component of RFID is RFID labels, antenna, and transceivers. In most of the recent innovations applications of the antenna, the smaller size of the antenna is demanded in every section and hence fascinates many antenna designers towards the miniaturizing of the antennas. To achieve the reduction in the size of the antenna, many different methods are implemented such as a meander line and fractal design or by implementing a specific higher permittivity substrate that favors the size reduction of the antenna by enhancing the actual field power in the higher permittivity areas. The radiation of the antenna is increased as a result of the lessening the size of the antenna due to the unbalanced distribution of the current on the smaller ground plane and the input connector part.

Due to the limitation in the RFID card size, nowadays, the researchers are working towards the antenna scaling down, hence in the proposed method Hilbert fractal hypothesis is used to accomplish the size reduction objective measurement of tag antenna. In addition, the impedance matching of the antenna will help improve antenna efficiency and productivity [13]. However, the IC chip input impedance is found to vary from manufacturer to manufacturer. According to the design rule hypothesis, the demanded tag antenna should directly match the input impedance of the IC chip. This is accomplished by balancing flexibly the input impedance of the tag antenna with the matched impedance of the antenna which is achieved by applying the meander line strategy. The simulation results demonstrate that tuning of the high line can estimate the input impedance of the tag antenna to change promptly. In this section, the microstrip meander line 3D antenna operating at 2.45 GHz for IoT condition is highlighted which aims at overcoming some of the drawbacks distinguished in conventional antennas which are nowadays utilized in certain applications. CST Microwave Studio is used to design and simulate the design processes of such an antenna to obtain the best performance of the antenna.

2.3.1. Design and the Analysis Approach

Figs. (**10** to **14**) elaborate the antenna design along with the antenna structure which is having a dimension of 21.6mm x 21.6mm mounted in the middle of the rectangular ground plane. Initially, λ/4 determines the antenna size which is determined as:

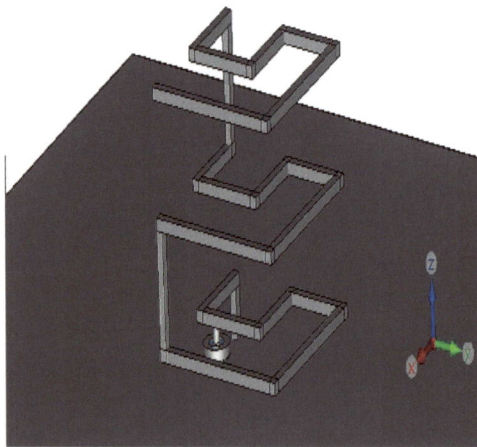

Fig. (10). Top View of Perspective 3D Antenna [14].

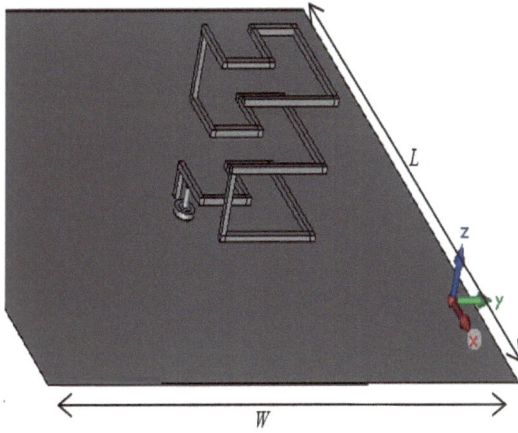

Fig. (11). The proposed 3D Antenna [14].

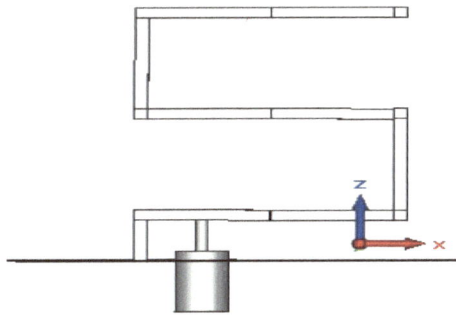

Fig. (12). Side View of the 3D Antenna [14].

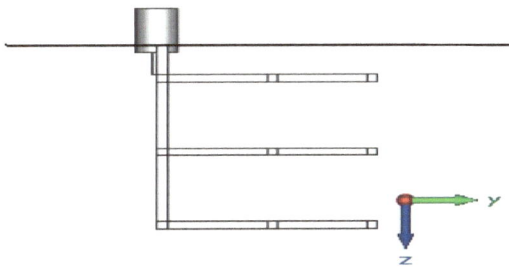

Fig. (13). Bottom View of 3D Antenna [14].

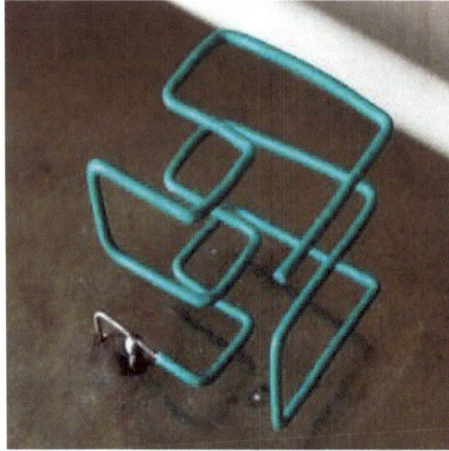

Fig. (14). Fabricated 3D miniature antenna [14].

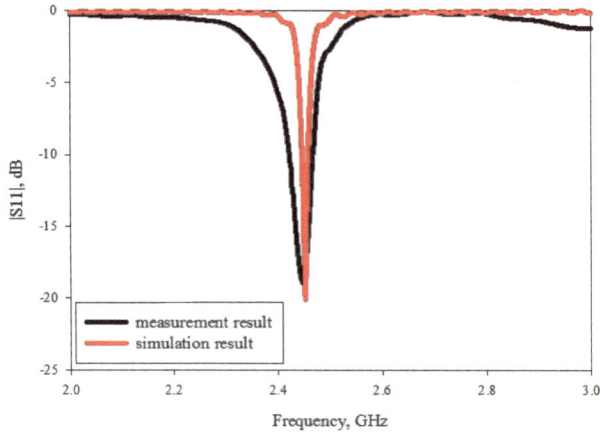

Fig. (15). S11| Antenna results [14].

$$\lambda = \frac{c}{f} \tag{8}$$

Where; λ=wave length, c=speed of light, f= frequency

$$\lambda = \frac{3x10^8}{2.44Ghz} = 0.1224\text{m} \tag{9}$$

By implementing the meander line technique in the 3D structure, the effective length of the antenna L2 is found to be 4.3mm which helps to reduce the antenna size to λ/30.

2.3.2. Simulation Results

Table **3** specifies the optimized antenna parameters dimension to fabricate the 3D antenna as illustrated by Fig. (**14**) to achieve the minimum return loss of -20 dB working at the frequency of 2.45 GHz. This designed 3D meander line antenna can cover the frequency range from 2.447 GHz to 2.459 GHz with a return loss of -10 dB. Consequently, the data transfer capacity of the designed antenna is established as 0.012 GHz which specifies the 48% at the center frequency.

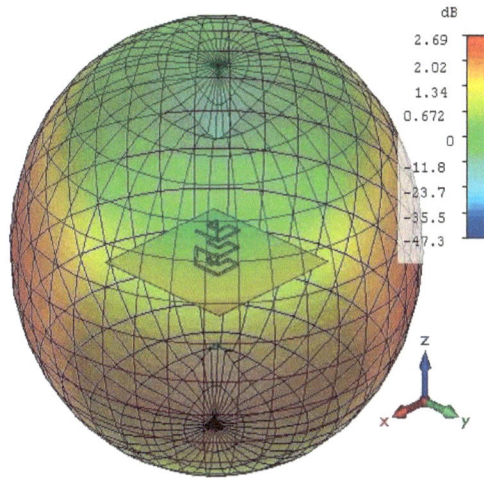

Fig. (16). 3D Antenna Radiation [14].

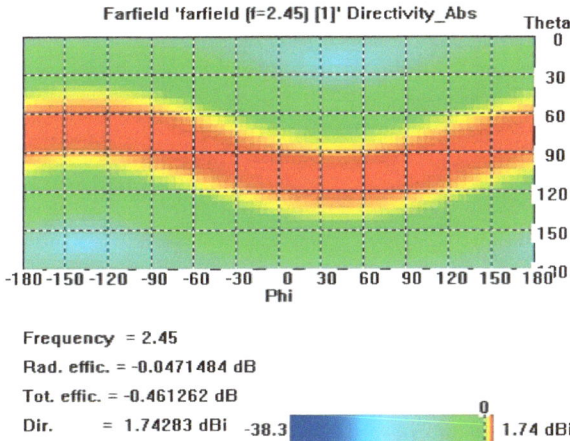

Fig. (17). 2D Antenna Radiation Pattern [14].

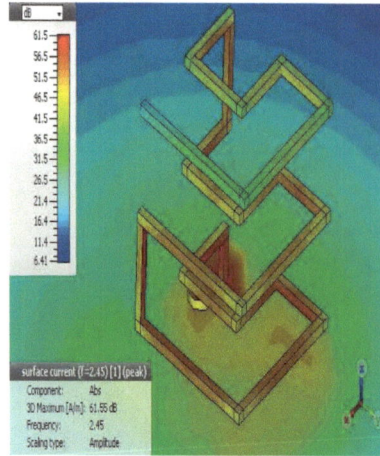

Fig. (18). Surface Current of the Antenna [14].

Table 3. The antenna size and parameter [14].

Parameters	Description	Value [mm]
L	Length of the ground plane	21.6
W	Width of the ground plane	21.6
L_1	Length of the antenna	4.3
L_2	Length of the antenna	2.15

The radiation pattern specifies the significant performance parameters of the antenna. Efficiency, antenna gain, and directivity are the performance parameters that can be recovered from the radiation pattern of the designed antenna. Figs. (**16** to **18**) illustrate the 3D radiation pattern as well as in 2D which significances that the radiation pattern describes an Omnidirectional radiation pattern and the far-field gain of the antenna is 2.69 dB at the frequency 2.45 GHz with the Main lobe direction of 103 degrees. The good performance of the antenna is obtained with the angular width at the 3 dB points of the main lobe is 95.3 degrees.

The power flux is considered entirely to determine the efficiency of the antenna which is estimated as -0.5 dB or 89%. The gain of the antenna signifies the main performance of the designed antenna which is consolidated with the antenna directivity and electrical efficiency whereas the directivity of the antenna represents the type of radiation or directional pattern. The directivity of this antenna is 1.28 dB at 2.45 GHz as shown in Fig. (**15**).

3. DESIGN CHALLENGES OF ANTENNA FOR IOT APPLICATIONS

The mobile phone or any IOT application uses an antenna as the basic component for connecting the various devices. However, the selection of the appropriate antenna for the specific application plays a key role in the design challenge. Various radio bands are utilized by wireless devices and the selection of reliable radio links is significant. The appropriate size and shape of the antenna along with its appropriate placement in the mobile and IoT devices play a key factor to enhance the antenna performance. Alone mobile phones contain different 4 to 13 antennas including 4 radios for transmission and reception of the data. The 4 radios include are mobile, Wi-Fi, Bluetooth, and GPS. Some of the other telephones may have additional radios: 802.15.4 (930 MHz and lower), FM radio, and magnetic Near-Field Communication (NFC).

However, the IoT remote devices contain one or two antennas whose size is much smaller than the antennas included in the mobile phone. For instance, smart-watches, health and fitness monitoring systems, skin-moisture monitors, vibration-sensors, broadcast chips in retail store-products, and numerous other applications that require the radio link for transmitting the data to a mobile phone or the Internet.

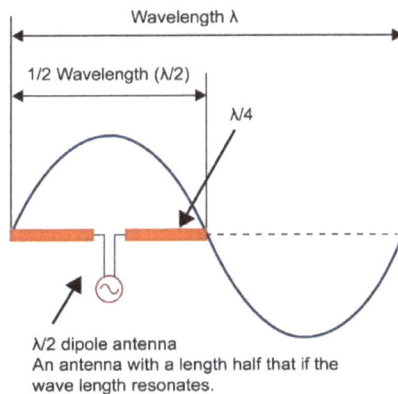

Fig. (19). Wavelength optimization [14].

3.1. Selection of the Appropriate Size of the Antenna

For the given space limitation in the mobile phone or IoT devices, the key challenge arises in the design of the antenna is, "How small can an antenna be and still it functions admirably?". According to the experts, the appropriate answer to the question is the one-fourth wavelength of the antenna. However, half the wavelength may be the ideal wavelength for the design of the antenna as shown in

Fig. (**19**) which appears as pure 73-ohm resistance in radio hardware which maybe even be reduced to one-fourth wavelength nearly 50 ohms resistance, and even though it works admirably [14].

Moreover, various tricks can be followed for the decrement of the size of the antenna, for example, utilizing the ground-plane of the circuit-board or by using the zig-zag-shaped antenna on the board which results in the reduction of the antenna size. These tricks may resemble the performance close to the effective length of one-fourth wavelength. However, an antenna size smaller than the one-fourth wavelength even works at the cost of a decrease in the signal quality accomplished with the area of the antenna and shrinkage in the accessible frequency bandwidth.

So what is the wavelength of all these radio bands? Two straightforward guidelines can help:

1. The frequency equals 300 MHz resembles a 1m radio signal wavelength (however, for America, the radio signal resembles 1ft for the frequency of 1 GHz).
2. As the wavelength of the radio signal varies inversely with the change in the frequency and hence it decreases with the increase in the frequency. Hence at 2 GHz frequency, the wavelength approximates half afoot.

By ignoring the NFC frequency (representing the magnetic-field-link), the radio signals operating at 88 MHz (which represents the FM waves) can work just at 6 GHz (representing Wi-Fi high band). Mobile phones work with high performance at the radio signals having the frequency band of 800 MHz to 5.5 MHz which results in the reduction of the wavelength from 15 to 2 in. The one-fourth wavelength of such radio signal ranges from 4 in down to 0.5 in. As a result, we can approximate that 5.5 GHz is the perfect match for the mobile phone, however, the antennas do not work properly at such a limited band. Thus, the wavelength of the FM radio band approximates around 10 ft and the tiny antenna used for its implementation will resemble terrible efficiency.

3.2. Critical Antenna Performance

Due to the size issues of the antenna used in mobile phones or IoT devices, the coverage of such antennas is limited as the radio signal drops at the cost of the square the distance of the separation, specially one-fourth of power of the distance due to any signal fluctuations or blockage and delay due to the spread of the signal from multiple reflections. If half of the radiated power is input to the antenna due to its size issues, then the usable coverage range of the device reduces by 15-30%. As a consequence, another big challenge faced by the mobile phone or the IoT device is finding the place for the antenna, particularly when the number of the different antennas are fighting for the useable areas.

3.3. Antennas Used in the Mobile Phones

Previously the trend of using small whip telescoping-antenna along with the mobile phone has become a distant memory. The antennas used in the mobile phones are small chip-antenna having 0.3 sq either mounted or imprinted on the circuit board or within the cover to have a reliable connection by using the board to the cover.

A patch antenna is one of the best-printed antennae is popularly used as it occupies less space as compared to the dipole antenna. It uses the ground plane of the board for the radiations.

A case of an iPhone mobile radio antenna is illustrated as follows:

As shown in Fig. (**20**), the radiating segment length is about 0.75 in and hence the effective length of the antenna segment is doubled by about 1.5 in using 869MHz uplink frequency with an effective wavelength of 0.11. This results in the reduction 0f antenna gain by 25-50%.

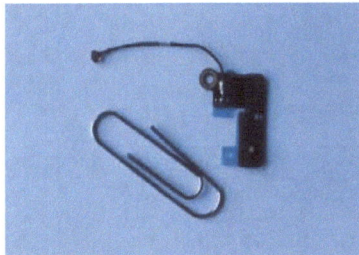

Fig. (20). iPhone antenna [14].

The radiating segment is about 0.75 in. and, hence the effective length of such segment is double that, or about 1.5 in. At the least mobile phone uplink frequency of 869 MHz, the effective length is about 0.11 wavelengths. This will lose in antenna gain, however, it is not sufficiently huge to cause an enormous drop in effective range.

The interference of the signals between multiple antennas in the mobile phone configuration is caused which becomes complicated as it depends more on electromagnetic simulation programs by which the antenna is modeled as well as it also included the impact of the conductor and the dielectric material used and exceptionally presence of the human body nearer to the antenna.

3.4. Antenna Used in IoT

These days, radio links are not only used in the mobile phone but also found applications including smart-watches,health-appliances, remote cameras, thermometers, home controlling appliances, *etc.,* along with Internet connections. The radios used in the IoT devices utilize similar principles and frequency bands as the mobile phone radios. The new devices in the IoT system face the issue that the antenna implemented into the device is allotted with less space as compared to that implemented into the mobile phones. As a result, an exceptionally small package assures that an apparel tag, a vibration screen, or a noise detector must include in it. Thus, the antennas used in smaller devices that are connected within the IoT system face the space limitation challenge [15]. Assume that the antennas are operating within the Wi-Fi band range of 5.5 GHz (where the 1cm antennas resemble 0.2 frequencies), then all the antennas will be operating inefficiently embedded into the tiny packages of sensors, as a consequence, the designer of such the smaller devices will occupy less range as compared to that of the mobile phone for the same power. However, if the enclosures are sufficiently small then the 1 GHz frequency bands cannot be utilized which allows the 2.4 GHz band (for Bluetooth, Zigbee, and Wi-Fi) and the 5.5 GHz band (Wi-Fi) to propose norms at the frequencies high to 60 GHz (which resembles the wavelength of 5mm) but beyond this frequencies may allow complex antennas with more efficiency.

4. CURRENT TRENDS IN THE DESIGN OF ANTENNAS FOR IOT APPLICATIONS

The solution to the required IoT hardware, first of all, requires sensors that will sense the surrounding environmental conditions then to monitor the output of the sensor, a remote dashboard is required that will display it in a clearer and conceivable form. Additional to this a serving and routing device is required which will perform the key task of the system by detecting specific conditions and taking actions accordingly. The communication between the devices and the dashboard should be secured. The sensors which can be commonly used in the IoT systems are accelerometers, temperature-sensors, magnetometers, proximity sensors, gyroscopes, image-sensors, acoustic-sensors, light-sensors, pressure-sensors, gas RFID-sensors, humidity-sensors & micro flow-sensors. Currently, many wearable devices like smart-watches, shoes, and 3D glasses are the best smart solution examples. The 3D glasses simply adjust the TV brightness and the contrast level according to our eyes and the smart-watches helps to track human daily activities and hence their fitness.

According to the research survey, it was concluded that the most important device which has tremendously contributed to IoT is cell phones contributing to revolutionizing the technology world. Mobile or Cell phones are embedded with many applications and the sensors which stores and reveals the information about its user. It has the advantage of elaborating Geo-location information and an embedded sensor that is capable of sensing and tracing light conditions, the orientation of your device, and a lot more information. It also contains many connectivity options like Wi-Fi, Bluetooth, and cells that help them to communicate with other devices [16].

The IoT applications are nowadays found many applications across all industries & market which has a multitude of expansion over various industries. The IoT devices span are popular overall groups of users, from those who are trying to reduce & conserve energy in their home to large organizations that want to improve their business operations. IoT has not only proved itself useful in optimizing critical applications in many organizations but also has boosted the concept of advanced automation which we have imagined a decade before.

Nowadays, the new IoT hardware system precedes the advancement domain in all aspects which started with mobile phones. To start with, the mobile phone towers comprise the mobile network associated with one side of the mobile phone and the Internet associated with it on the other side. Talking about Wi-Fi, the PCs are connected to the Internet through access points (hotspots) either at home or any public place. Thereafter the Wi-Fi radios are implemented into mobile phones which can be connected to the internet through mobile phone networks or the hotspots of the Wi-Fi. These days, the wireless diffuse networks, known as Mesh systems, are developing with numerous small access points and implemented in the applications in the field of industrial as well as in-home automation. Thus, the final product results empowered with the small radios are practically cost-efficient and access instantly to the Internet or other devices in the System. The one thing to appreciate is that the antennas are indiscriminate which means they do not care about the energy of the signal but mainly concern about the frequency and its strength. The antenna is designed in such a way that the modulation signal should carry the real data and should be able to recover the data from the backend in a sympathetic way, thus the antenna should follow the theory of reciprocity an antenna will behave in the same way when it is both receiving and transmitting. This means that it does not matter if the device is a transmitter, a receiver, or both, the antenna design will be the same. The antenna employed in the small devices of the IoT system is classed as either embedded, by mounting it directly to the PCB and connected using coaxial cable. The antennas are mounted outside the main enclosure or the building. The design of the antenna implemented into the IoT system considers several criteria like date rate, frequencies, and the range of the wireless connection which will enhance the power levels of the system.

For two decades, Molex an antenna design company has been designing, manufacturing, and supplying standard and custom integrated antennas around the world for mobile ad IoT applications whose product portfolio will include all the major protocols as well as proprietary solutions, with a focus on embedded (PCB mounted) and internal cabled antennas, when performance requirements may dictate their use [17, 18].

The antenna which is embedded is extremely smaller in size and manufactured from ceramic and plastic as shown in Fig. (**21**) which may weigh just a few grams. Cabled antennas are made of rigid and flexible stamped metal PCB which makes it simple to find the most appropriate antenna design, whatever the application. Even though they can be made very small and compact, embedded antennas still come with design requirements that, if followed, will help ensure that the antenna performs as the specification demands. Many factors, such as the size of the PCB, its shape, and where the antenna is mounted on the PCB can all impact how the antenna performs.

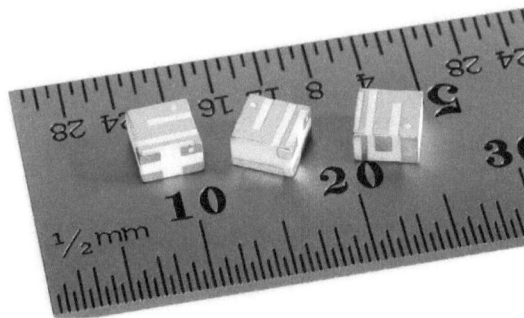

Fig. (21). Antennas can be designed and manufactured to meet the most demanding space constraints.

Perhaps most importantly, the application specification provides an estimate of the antenna's operation and how this will vary based on its location on a PCB, taking into account its proximity to other components and features such as batteries, metal shields, and the cables itself for cabled antennas as shown in Fig. (**22**). This application specification documents provide essential guidelines that system designers should read to understand how to best engineer the antenna into each system.

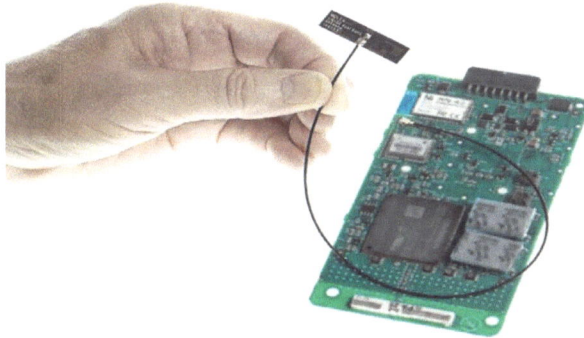

Fig. (22). Design options include PCB-mounted and remotely located. Often, the application will dictate the most appropriate approach.

In most cases, the information provided in the application specification has been simplified to make it more easily accessible to engineers with varying levels of RF expertise, and of course, Molex can also provide access to an expert if and when required. In many instances, it is advised that design teams placing antennas into their systems seek assistance and guidance from an RF expert. This can save time when it comes to understanding what additional measures are needed to boost antenna performance and quantitatively measuring antenna performance in a system.

To support its customers, Molex has also invested in antenna measurement technology, including field-scanning chambers to support product development, up to frequencies used in 5G transmissions. When used alongside its advanced simulation capabilities, this enables Molex to develop antennas that deliver high performance while still complying with SAR (Specific Absorption Rate) specifications. Once designed, Molex can also manufacture custom antennas, to meet the customer's production needs. Molex not only designs antennas; its RF engineers are experts and are ready to assist customers' design engineers investigate and document the performance of antennas in a system.

CONCLUSION

Internet of Things is another upheaval of the Internet and it is a key exploration subject for a specialist in embedded software engineering and information technology because of its different areas of utilization, heterogeneous combination of different communication and embedded innovation in its design. Today, a considerable lot of these obstacles have been addressed. The size and cost of wireless radios have dropped enormously. IPv6 permits us to allocate a correspondence address to billions of devices. Devices organizations are incorporating Wi-Fi and cell remote availability into a wide scope of devices.

Mobile information inclusion has improved fundamentally with a huge number contribution broadband speeds. Battery innovation has been improved and solar-powered re-energizing has been incorporated into various devices. There will be billions of objects interfacing with the network in the following several years. For example, Cisco's Internet of Things Group (IOTG) predicts there will be more than 50 billion associated devices by 2020.

In this chapter, we have discussed the different antennas and their implementation applicable in IoT systems. Firstly, we discussed the designing of a multi-band IoT antenna having a notch bandwidth of 10 MHz, 40 MHz, and 10 MHz resembling the approximate resonate frequencies 825 MHz, 870 MHz, and 1195 MHz by simulating in the HFSS programming environment. At the resonant frequencies, the return loss is optimized to -20 dB which is far better than the minimum requirement of -10dB. Subsequently, the specifications of the application are fulfilled by the performance of the antenna [7].

Secondly, we also discussed that the antenna designed utilizing RT/duroid is more efficient when compared with the designed antenna utilizing FR4 epoxy. For the designed rectangular-patch along with a rectangular-slot and the ground structure punctured with a circle, the impedance matching has been poor, and consequently with FR-4 ($\varepsilon r = 4.4$) as substrate, there is hardly a frequency at which the antenna will, in general, tend to work [13].

We also discussed the 3D miniaturized antenna used for the IoT application. The simulation results show that the power flux is considered entirely to determine the efficiency of the antenna which is estimated as -0.5 dB or 89%. The gain of the antenna signifies the main performance of the designed antenna which is consolidated with the antenna directivity and electrical efficiency whereas the directivity of the antenna represents the type of radiation or directional pattern. The directivity of this antenna is 1.28 dB at 2.45 GHz.

We concluded that radio links used in the IoT devices utilize similar principles and frequency bands as the mobile phone radios. It is also concluded that the antenna implemented into the IoT device is allotted with less space as compared withthat implemented into the mobile phones. As a result, an exceptionally small package assures that an apparel tag, a vibration screen, or a noise detector must include in it. Thus, the antennas used in smaller devices that are connected within the IoT system face the space limitation challenge [15]. According to the research survey, it was concluded that the most important device which has tremendously contributed to IoT is cell phones contributing to revolutionizing the technological world. Mobile or Cell phones are embedded with many applications and the sensors which stores and reveals the information about its user. It has the advantage of elaborating Geo-location information and an embedded sensor that is capable of sensing and tracing light conditions, the orientation of your device, and a lot more information. It also contains options related to connectivity like Wi-Fi, Bluetooth, and cells that help them to communicate with other devices [16].

It is also concluded that the antennas are indiscriminate which means they do not care about the energy of the signal but mainly concern about the frequency and its strength. The antenna is designed in such a way that the modulation signal should carry the real data and should be able to recover the data from the backend in a sympathetic way, thus the antenna should follow the theory of reciprocity an antenna will behave in the same way when it is both receiving and transmitting. This means that it doesn't matter if the device is a transmitter, a receiver, or both, the antenna design will be the same.

CONSENT FOR PUBLICATION

Not Applicable.

CONFLICT OF INTEREST

The author confirms that this chapter contents have no conflict of interest.

ACKNOWLEDGEMENT

Declared none.

REFERENCES

[1]　C.A. Balanis, *Antenna Theory: Analysis and Design.* Wiley: Hoboken, NJ, USA, 2005.

[2]　"Harisha, D Sai Prashanth, Bharath Bhushan SN," Performance Analysis of Circular Shaped Patch Antenna for IoT Application Device", *IJCSMC,* vol. 5, no. Issue. 10, pp. 48-52, 2016.

[3]　J. Pei, A-G. Wang, S. Gao, and W. Leng, "Miniaturized triple-band antenna with a defected ground plane for WLAN/WiMAX applica- tions", *IEEE Antennas Wirel. Propag. Lett.,* vol. 10, p. 298301, 2011.

[4]　I. Dioum, A. Diallo, S.M. Farssi, and C. Luxey, "A novel compact dual-band LTE antenna-system for MIMO operation", *IEEE Trans. Antenn. Propag.,* vol. 62, no. 4, p. 22912296, 2014. [http://dx.doi.org/10.1109/TAP.2014.2301151]

[5]　Y. Dong, H. Toyao, and T. Itoh, "Design and characterization of miniatur- ized patch antennas loaded with complementary split-ring res- onators", *IEEE Trans. Antenn. Propag.,* vol. 60, no. 2, p. 772785, 2012. [http://dx.doi.org/10.1109/TAP.2011.2173120]

[6]　Y.T. Lo, and S.W. Lee, *Antenna handbook: Antenna theory.* vol. Vol. 2. International Thompson Publishing, Inc, 1993.

[7]　J.C. Narayana Swamy, Multiband Microstrip Patch Antenna for IoT Applications",International Journal of Innovative Technology and Exploring Engineering (IJITEE) ISSN: 2278-3075, Volume-9

Issue-1, November 2019.,

[8]   P. Beigi, and J. Nourinia, *A novel printed antenna with square spiral structure for WIMAX and WLAN applications.* vol. Vol. 30. IEEE, 2007.

[9]   Y.I. Abdulraheem, "Design of frequency reconfigurable multiband compact antenna using two pin diodes for WLAN/WIMAX Applications" IET microwaves, volume 11, 2017.,

[10]   P. Beigi, and P. Mohammadi, *Bandwidth enhancement of monopole antenna with DGS for SHF and reconfigurable structure for WIMAX, WLAN and c-band applications.* vol. Vol. 12. SISSA, 2017.

[11]   M. Shrama, Y.K. Awasthi, and H. Singh, "Planar high rejection dual band-notch UWB antenna with X& Ku- bands a Wireless Applications", *Int. J. Microw. Wirel. Technol.,* 2017. [http://dx.doi.org/10.1017/S1759078717000393]

[12]   B. Srikanth Deepak, R. Naveen Kumar, M. Madhavi, P. Avinash, and K. Yaswanth, "Design of A Handy Tripleband Micro- Strip Patch Antenna for Satellite Based Iot Applications", *International Journal of Innovative Technology and Exploring Engineering (IJITEE),* vol. 8, no. 6, pp. 2278-3075, 2019.

[13]   V.D. Hunt, A. Puglia, and M. Puglia, *RFID: A Guide of Radio Frequency Identification.* John Wiley & Sons, 2007. [http://dx.doi.org/10.1002/0470112255]

[14]   A. Nasir Mohamed, S.N. Azemi, S.A. Suhaimi, and A.A.M. Ezanuddin, 3D Miniature Antenna Design for RFID Applications in IoT Environment",MATEC Web of Conference, [http://dx.doi.org/10.1051/matecconf/20179701092]

[15]   Z. Katbay, S. Sadek, R. Lababidi, A. Perennec, and M. Le Roy, "In New Circuits and Systems Conference (NEWCAS)", *IEEE 13ᵗʰ International,* 2015.

[16]   P.K. Malik, "Industrial Internet of Things and its Applications in Industry 4.0: State of The Art", *Computer Communication,* vol. 166, Elsevier, pp. 125-139, 2021. ISSN 0140-3664. [http://dx.doi.org/10.1016/j.comcom.2020.11.016]

[17]   P.K. Malik, D.S. Wadhwa, and J.S. Khinda, "A Survey of Device to Device and Cooperative Communication for the Future Cellular Networks", *Int. J. Wirel. Inf. Netw.,* 2020. [http://dx.doi.org/10.1007/s10776-020-00482-8]

[18]   N. Shaik, and P.K. Malik, "A Retrospection of Channel Estimation Techniques for 5G Wireless Communications: Opportunities and Challenges", *International Journal of Advanced Science and Technology,* vol. 29, no. 05, pp. 8469-8479, 2020.

CHAPTER 6

Use of Smart-Antenna in Mobile Communication

Arghya Majumder[1,*]

[1] *Department of Electrical and Electronics Engineering, Lovely Professional University, Punjab, India*

Abstract: Smart antenna is one of the latest inventions in science and technology, which facilitates the communication process in a better way. There are a lot of features in smart antennas. It has been very useful to track the desired antenna's location by calculating the beam formation. Due to the implementation of smart signal processing algorithm, we can use the spectrum efficiently. There is a reduction of cost in establishing new wireless networks. It also gives more reliable quality of service. In this chapter, we will talk about the benefits and the most recent advances in smart antenna transceiver architecture. By researching the smart antenna techniques, it is observed that the success of the smart antenna lies on two major factors:

i) Top-down compatibility: For next generation communication system, the features of smart antenna need to be preferred.

ii) Bottom-up feasibility: According to critical parameters, smart antenna technique performs realistically.

At the end we will discuss market trends, future projections, and the expected financial impact of smart antenna systems deployment.

Keywords: Antenna Performance, Internet of things, Microstrip antenna, Wireless Communication.

1. INTRODUCTION

The array of many smaller antennas, which are connected using intelligent signal processing software, is known as a smart antenna. Smart antenna system consists of multiple antenna elements at the transmitting and the receiving end of the communication link. When the spatial dimension is reduced, the capacity of wireless network is increased as the link quality is improved by the mitigation of a number of disabilities of mobile communication. We can do it by increasing the data rate through the simultaneous transmission of multiple streams.

---

[*] **Corresponding author Arghya Majumder:** Department of Electrical and Electronics Engineering, Lovely Professional University, Punjab, India; E-mail: arghya.majumder99@gmail.com

Praveen Kumar Malik, Pradeep Kumar, Sachin Kumar and Dushyant Kumar Singh (Eds.)

Smart antennas are also termed as multiple antennas, adaptive array antennas *etc*. It increases the efficiency in digital wireless communication systems. It works based on the diversity effect source and destination of the wireless system. We also use the smart antenna to calculate the beam forming vectors and the direction of arrival of the signal.

In upcoming days, every wireless system will come up with built-in smart antenna technology that will provide a better service quality, reduce the cost of implementation of new wireless network, create a beneficial impact on efficient use of spectrum and reconfigure the disputes by smart realization of the signals.

In this chapter, we will briefly discuss the latest trends and the features of the smart antenna for the upcoming generation. We will also discuss the major challenges in mobile radio communication and the improvement in the performance of mobile communication systems by the effective use of smart antennas.

1.1. Types of Smart Antenna Systems

There are few common terms like digital beam formation, adaptive antenna system, phased array, intelligent antenna, spatial processing *etc*., that support various work function of smart antenna. Given below are the two major categories of smart antennas which are differentiated by the transmission strategy:

1. Switched Beam—It is having a finite number of fixed, predefined patterns or combining strategies.
2. Adaptive Array—It has an infinite number of patterns that are adjusted in real time.

1.2. Major Challenges in Mobile Communication System

Among the few major challenges in mobile communication systems, multipath propagation is one of them. The creation of multiple signal directions between the transmitter and the receiver due to the reflection of the transmitted signal by physical obstacles is called Multipath propagation.

Address migration is another challenge in mobile communication system. Today's networking is not designated for dynamically changing addresses. Active network connections cannot be moved to a new address. Human intervention is usually required to coordinate the use of address.

Huge power consumption is also considered as a major challenge of mobile communication. Huge power consumption is realized when the mobile is travelled from one address to another address, as the mobile aims to connect its network for one IP to another IP. For this reason, the heat is also dissipated.

On the other hand, co-channel interference is a major limiting factor on the capacity of wireless systems, resulting in the reuse of the available network resources by a number of users.

1.3. Improvement the Performance of Mobile Communication System by Using Smart Antenna

The link quality can be improved in smart antenna by combating the effects of multipath propagation or constructively exploiting the different directions, and increasing the capacity by preventing interference and allowing transmission of different data streams from different antennas. The benefits of smart antennas are given below:

1. Expanded the area of coverage: Sometimes coherent signals are combined in the receiver terminal of antenna. Due to this, the beamforming gain is increased. When there are few number of antennas, the beamforming gain is also increased as both are proportional to each other.
2. Reliability: When the smart antennas with spatial domain are sampled with each other, then a diversity of signals is obtained. The effective fluctuation of the signal is reduced due to the presence of the sampled smart antennas through fading signal parts.
3. Low power requirement: This technology performs under low power as this technology provides the features to the wanted users only at the required time and application basis.
4. Multiple path rejection: Smart antenna works on higher bit rates without the use of equalizer. Due to this, the amount of multipath is decreased as effective delay spread of the channel is reduced.
5. Increased spectral efficiency: Smart antenna uses a multiple access scheme, called Space division multiple access (SDMA) which increases the spectral efficiency. SDMA increase the spectral efficiency, which increase the date rate.

2. SMART ANTENNA STRATEGIES FOR MOBILE COMMUNICATION

We expect theat next generation mobile communication may have more reliability, good service quality, more bit rates and less signal fluctuations. The smart antenna provides the requirements by less propagation delay, flexible user mobility, reconfiguring antenna's location or IP address, and less interference. Smart antenna design has achieved the advancements in the next generation mobile communication.

Few strategies of smart antenna for next generation mobile communication are given below:

2.1. Cross-layer Optimization

Physical layer, link and the network layers develops the Smart antenna technique by combining their parameters. The link parameters are medium access control, data link control, scheduling, *etc.* And the layer networks refer to radio resource management, routing, *etc.* An isolated layer proves its efficiency after the evaluation of the performance in the higher layer. The signal is transmitted among the functionalities residing in different OSI layers. The OSI layers are further classified into: CSI, QoS parameters and physical layer resources.

2.2. Multi-user Diversity

Smart antenna is specially designed for multiple users by providing the channels. There are higher chances of completing a successful transmission due to this multiplexing. Opportunistic approaches go beyond the physical layer and exploit multi-user diversity as a complement to code, time, frequency, or space diversity. The smart antenna designed based on medium access control is forced to reduce the collision avoiding paradigm, it is evolution toward multi-user schemes, thus reinforcing the need for cross-layer designs.

2.3. Performance

The application of smart antennas in future mobile communication systems depends on two main approaches:

For next generation communication system, the features of smart antenna need to be preferred mostly.

According to critical parameters, smart antenna technique performs realistic.

2.4. System Perspective

By using smart antenna in mobile communication, the communication system gets easier in terms of traffic behaviour of signals, multiple access methods and beamforming. We all know that the wireless channels fluctuate in both time and position. The fluctuation may happen due to beamforming. To remove this fluctuation multi-user diversity is introduced. There is a process, called scheduling algorithm, which schedule the communications to a channel which is more suitable and more reliable than the other channels.

CONCLUSION

The antennas with intelligence and smart signal processing function is called smart antenna. Multiple-input-multiple-output system is having a characteristic like smart antenna. The smart antenna is a system which works depending upon the diversity of transmitter, receiver or both. Beamforming is a vital term in this Smart antenna technique. It is obtained by adaptation techniques (like RLS). In smart antenna the radiation patterns may vary with unchanged mechanical parameters. Basically there are two types of smart antenna: switched beam smart antenna and adaptive array smart antenna. We can calculate the beam direction by direction of arrival method, which is known as DOA method.

CONSENT FOR PUBLICATION

Not Applicable.

CONFLICT OF INTEREST

The author confirms that this chapter contents have no conflict of interest.

ACKNOWLEDGEMENT

Declared none.

REFERENCES

[1]   Alexiou A., and Haardt M., "Smart Antenna Technologies for Future", *IEEE communications Magazine,* vol. 42, no. 9, pp. 90-97.

[2]   A. K. Gupta, "Proceedings of 2nd National Conference on Challenges & Opportunities in Information Technology (COIT-2008) RIMT-IET", *Challenge in Mobile Computing,* Mandi Gobindgarh, 2008.

[3]   B. Rahul, and A.B. Mannade, "Challenges of Mobile Computing: An Overview", *Int. J. Adv. Res. Comput. Commun. Eng,* 2013.

[4]   P.K. Malik, D.S. Wadhwa, and J.S. Khinda, "A Survey of Device to Device and Cooperative Communication for the Future Cellular Networks", *Int. J. Wirel. Inf. Netw,* 2020.
[http://dx.doi.org/10.1007/s10776-020-00482-8]

[5]   N. Shaik, and P.K. Malik, "A Retrospection of Channel Estimation Techniques for 5G Wireless Communications: Opportunities and Challenges", *International Journal of Advanced Science and Technology,* vol. 29, no. 05, pp. 8469-8479, 2020.

[6]   Kumar Malik Praveen, and Singh Madam, "Multiple Bandwidth Design of Micro strip Antenna for Future Wireless Communication", *International Journal of Recent Technology and Engineering,* vol. 8, no. 2, pp. 5135-5138, 2019. ISSN: 2277-3878.
[http://dx.doi.org/10.35940/ijrte.B2871.078219]

[7]   V. Arora, and P.K. Malik, "Analysis and Synthesis of Performance Parameter of Rectangular Patch Antenna", *Proceedings of First International Conference on Computing, Communications, and Cyber-Security (IC4S 2019). Lecture Notes in Networks and Systems,* vol. 121, Springer: Singapore, 2019.
[http://dx.doi.org/10.1007/978-981-15-3369-3_12]

SUBJECT INDEX

A

B

C

miniaturizing 65
modern printed circuit 48
modern telecommunications 51
multiband MIMO Antenna 41
wearable computer device 52
wireless communication 40
Telecommunication industry 11
Textile 51, 52
  applications 51
  materials 52
  systems, intelligent 52
Trading costs 32
Traffic behaviour 87
Transfer data 58
Transmission 20, 45, 47, 49, 51, 52, 74, 80,
    85, 86, 87
  automated supply-chain data 49
  efficient data 49
  knowledge 52
  solar space 47
  strategy 85
  wireless power 47, 51
TV 46, 77
  brightness 77
  broadcasts 46

U

Uniform geometrical diffraction theory 21

V

Vehicle 1, 34
  collision avoidance system 1
  communication 34
  guidance system 1
  manufactures 34
Versatile properties 64
Voltage standing wave proportion 65

W

Wavelength 74, 76
  effective 76

ideal 74
Waves, high-frequency 45
Wi-Fi 46, 77, 78
  band range 77
  radios 78
  routers 46
  routers and wireless endpoints 46
Wireless 1, 11, 22, 23, 40, 41, 45, 47, 49, 74,
    79, 80, 86, 87
  applications, handheld 23
  channels fluctuate 87
  connection 79
  data 45
  devices 74
  local area network (WLAN) 1, 11, 22, 40,
    47
  network applications 41
  radios 80
  service 49
  systems 85, 86
Wireless Communication 8, 11, 12, 22, 23, 25,
    41, 48, 57, 84
  applications 8, 12, 23
  areas 41
  devices 22
  industry 25
Wireless communication systems 8, 41, 85
  digital 85

www.ingramcontent.com/pod-product-compliance
Lightning Source LLC
Chambersburg PA
CBHW041720210326
41598CB00007B/720